Amphibians and Reptiles
in Minnesota

Also Published by the University of Minnesota Press

The Mammals of Minnesota
Evan B. Hazard
Illustrations by Nan Kane

Northern Pike: Ecology, Conservation, and Management History
Rodney B. Pierce
Minnesota Department of Natural Resources

Northland Wildflowers: The Comprehensive Guide to the Minnesota Region
John B. Moyle and Evelyn W. Moyle
Photography by John Gregor

Native Orchids of Minnesota
Welby R. Smith
Minnesota Department of Natural Resources

Trees and Shrubs of Minnesota
Welby R. Smith
Minnesota Department of Natural Resources

AMPHIBIANS
AND
REPTILES
IN
MINNESOTA

John J. Moriarty and Carol D. Hall

Minnesota Department of Natural Resources

Foreword by Carrol L. Henderson

University of Minnesota Press

Minneapolis · London

The University of Minnesota Press gratefully acknowledges financial assistance provided by the Minnesota Department of Natural Resources for the publication of this book.

Partial funding for this book was provided by the Minnesota Environment and Natural Resources Trust Fund as recommended by the Legislative-Citizen Commission on Minnesota Resources (LCCMR).

Generous financial support toward the publication of this book was provided by the Minnesota Herpetological Society.

Some of the material in this book first appeared in *Amphibians and Reptiles Native to Minnesota* by Barney Oldfield and John J. Moriarty (Minneapolis: University of Minnesota Press, 1994).

Published by the University of Minnesota Press
111 Third Avenue South, Suite 290
Minneapolis, MN 55401-2520
http://www.upress.umn.edu

Library of Congress Cataloging-in-Publication Data
Moriarty, John J.
  Amphibians and reptiles in Minnesota / John J. Moriarty and Carol D. Hall, Minnesota
  Department of Natural Resources ; foreword by Carrol L. Henderson.
  Includes bibliographical references and index.
  ISBN 978-0-8166-9091-6 (pb. : alk. paper)
  1. Amphibians—Minnesota. 2. Reptiles—Minnesota. 3. Amphibians—Minnesota—Identification.
4. Reptiles—Minnesota—Identification. I. Hall, Carol D., 1957– II. Title.
  QL653.M55M67   2014
  597.809776—dc23

                                                                           2013039570

Printed in South Korea on acid-free paper

The University of Minnesota is an equal-opportunity educator and employer.

20 19 18 17 16 15 14   10 9 8 7 6 5 4 3 2 1

Contents

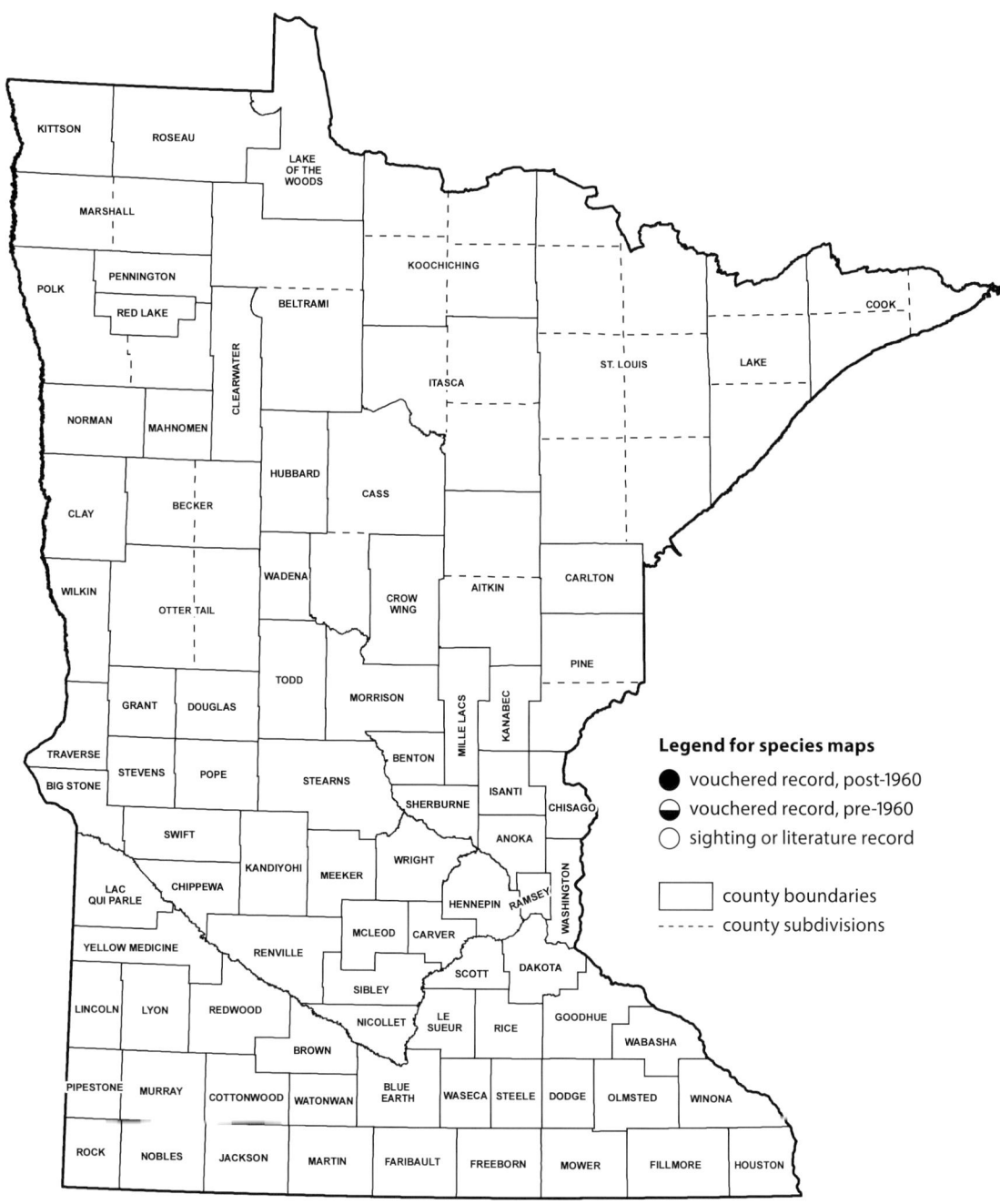

Minnesota county map

Foreword

Carrol L. Henderson

Frogs, toads, turtles, and snakes! What better way to introduce children to nature than with the wonderful array of wildlife that can be discovered literally in your own backyard, including reptiles and amphibians. For many people, that childhood interest in reptiles and amphibians grows into a lifelong interest in and concern for all nature.

There are many ways to enjoy these creatures. Some species, like the Spring Peeper, announce springtime with their calls. Northern Leopard Frogs provide infinite fascination for children exploring the shores of our lakes and marshes, and Western Painted Turtles are a constant shoreline companion for anglers and canoeists throughout the state.

Amphibians serve important purposes in genetic research, and the Kandiyohi variant of the Northern Leopard Frog has been used in cancer research. Reptiles and amphibians can incur deformities and other survival problems because of toxic or carcinogenic chemical pollutants in the environment. Long-lived reptiles, like the Snapping Turtle, are sometimes caught and eaten, but they can carry persistent environmental contaminants in their body for several decades. They are a continuing reminder that we need to pay attention to the health of reptile and amphibian populations as a part of responsible environmental stewardship for our state's lakes and rivers.

Minnesota is fortunate to have talented and experienced herpetologists and a very active Minnesota Herpetological Society, where members share their lifelong passion for those assorted scaly, slimy, slippery, crawling, and creeping creatures that comprise such an important part of our state's biological diversity.

The first book on our state's amphibians and reptiles, *The Reptiles and Amphibians of Minnesota,* was published in 1944 by Dr. Walter Breckenridge, noted ornithologist, photographer, author, herpetologist, and former director of the Bell Museum of Natural History at the University of Minnesota. Fifty years later, in 1994, Barney Oldfield and John Moriarty published a significantly updated book, *Amphibians and Reptiles Native to Minnesota,* through collaboration with the Nongame Wildlife Program in the Minnesota Department of Natural Resources (DNR). During the twenty years since the publication of that book, the Minnesota Biological Survey of the DNR has generated a huge amount of new information on the status, occurrence, and distribution of our reptiles and amphibians. During this era, analysis of the DNA of reptiles and amphibians has resulted in significant changes in organization of families and scientific names. Additional fieldwork by both amateur and professional herpetologists throughout the state has contributed considerably to our collective knowledge of our herpetological resources.

This book provides an opportunity to recognize the importance of our reptile and amphibian resources in Minnesota and to acknowledge the lifelong accomplishments of herpetologists John J. Moriarty and Carol D. Hall, who have written this book to share with you their knowledge and enthusiasm for the reptiles and amphibians of Minnesota.

Preface and Acknowledgments

It has been twenty years since *Amphibians and Reptiles Native to Minnesota* (Oldfield and Moriarty 1994) was published. The interest in and information on amphibians and reptiles increased greatly over that time. In particular, the natural history and distributional information grew steadily over the years. The Minnesota Biological Survey (MBS) visited most of the state's counties, gathering large numbers of distributional records. The Minnesota Herpetological Society increased its field trips and survey work. Finally, the general public provided a great deal of important information by way of sightings and other observations. We received many records from people who read the earlier edition of this book and recognized the frog or snake in their yard was not mapped or was doing something not described in the book.

Three additional species (Spotted Salamander, Four-toed Salamander, and Eastern Musk Turtle) have been found in the state, as well as one introduced species, Pond Slider. Tiger Salamander was split into two species, Eastern Tiger Salamander and Western Tiger Salamander. The taxonomy has changed tremendously, resulting in more than 40 percent of Minnesota species experiencing a change in genus or species names.

The interest in the conservation of reptiles and amphibians has grown. This can be seen in the large volunteer response to the Minnesota Frog and Toad Calling Survey and the public's concern about turtles crossing roads.

We have each been chasing and studying amphibians and reptiles in Minnesota for almost three decades. It is still exciting for us to catch a frog, help a snake off the road, or watch a turtle nesting in the spring. We hope our efforts in this book will inspire others to become excited about amphibians and reptiles in Minnesota.

A number of friends and colleagues helped collect information and take photographs and provided a sounding board for our ideas. Several also helped with *Amphibians and Reptiles Native to Minnesota*. We would like to acknowledge their continued support.

We first acknowledge Barney Oldfield for his work on the first edition and for his support of this book. Barney felt that his move to New Mexico more than ten years ago has kept him away from recent Minnesota herpetology.

Current and past employees of the DNR gave us distributional records and natural history information, especially Rich Baker, Jamie Edwards, Joan Galli, Lisa Gelvin-Innvaer, Maya Hamady, Erica Hoaglund, Krista Larson, Pam Perry, Ed Quinn, John Schladweiler, Konrad Schmidt, and folks from the Minnesota Biological Survey, who collected many of the distributional records: Liz Harper, Jeff LeClere, Gerda Nordquist, Kelly Lynch Pharis, and Christi Spak. Jared Cruz helped with data management. Many members of the Minnesota Herpetological Society and the public contributed, especially Noah Anderson,

Randy Blasus, Tony Gamble, Jim Gerholdt, Bruce Haig, Nancy Haig, Tom Jessen, Dan Keyler, Chris Smith, Bruce Brecke, Gary Casper, Phil Cochran, Steve Freedberg, Judy Helgen, Tim Lewis, Madeleine Linck, Mike Pappas, and Bill Souder.

The photographs in the book were provided by excellent photographers and colleagues. We thank Allen Blake Sheldon, Tony Gamble, Jim Gerholdt, Tom Jessen, Jeff LeClere, Barney Oldfield, Chris Smith, John Tester, and the DNR staff (Barb Delaney, Eric Halbur, Fred Harris, Michael Lee, Kelly Lynch Pharis, Gerda Nordquist, Kurt Rusterholz, Bernard Seitman, Christi Spak, Tim Whitfeld, and Scott Zager).

The introductory maps, distribution maps, and keys were expertly drafted by MBS graphics artist Tom Klein. Jeffrey Parmelee kindly allowed us to revise his larval amphibian keys.

Distributional records were collected by reviewing the amphibian and reptile collections at the Bell Museum of Natural History. We thank current curator Ken Kozak and past curator Andrew Simons and their curatorial assistants Tony Gamble, Ben Lowe, Amy Luxbacher, and Chris Smith for checking records and verifying specimens. Don Luce offered access to the Bell Museum's photography archives.

The preparation of this book was funded by the DNR's Nongame Wildlife Program. We acknowledge Carrol L. Henderson and Carmen Converse for their efforts and support with contract and funding negotiations for the book. Todd Orjala from the University of Minnesota Press provided support and encouragement.

Krista Larson and Tim Lewis read the entire manuscript and contributed excellent comments that improved the book.

Numerous herpetologists and biologists supported this and previous efforts. We thank David Hoppe, Jeff Lang, Robert McKinnell, John Tester, and all the others too numerous to name.

Most important, we thank our immediate families, Mike Hall, JoAnne Wetherell-Moriarty, and Caitlin and Daniel Moriarty, for their support and understanding during the past couple of years.

Introduction

Minnesota's amphibians and reptiles are an interesting and diverse group of vertebrates. There are 52 species native to the state, plus 1 introduced species, and their variety in form, size, and color is indeed remarkable. Minnesota has 22 species of amphibians including 8 salamanders and 14 anurans (frogs and toads). Of the 31 species of reptiles inhabiting the state, there are 11 turtles (one introduced), 3 lizards, and 17 snakes. Many of the species native to Minnesota are at the northern limits of their range. Frequently, a species range extends into the state along the Mississippi River valley; thus, the southeastern portion of the state offers the greatest abundance and diversity of amphibians and reptiles.

Several Minnesota species are habitat generalists and are relatively common throughout the state. The American Toad, Common Gartersnake, Painted Turtle, and Northern Leopard Frogs are examples. Others have special habitat requirements, and their distribution is limited to localized areas. Species such as the Plains Hog-nosed Snake, Wood Turtle, Common Five-lined Skink, and Mink Frog fit this category. A third group includes extremely rare species, such as the Lined Snake and Western Ratsnake, as their range just barely extends into Minnesota from neighboring states.

WHAT IS A HERP?

Herpetology is the discipline for the study of amphibians and reptiles, and a person who studies these animals is a herpetologist. *Herp* is a popular colloquial term that refers to either an amphibian or a reptile. Derivatives of the word herp that are in widespread use include *herper* (someone interested in amphibians and reptiles, normally field oriented), *herping* (searching for amphibians and reptiles), and *herp societies* (organizations devoted to the study and captive care of amphibians and reptiles).

Amphibians and reptiles are not close relatives, but placing these two classes of animals into one group for study is convenient for several reasons. Characteristics of animals in both classes include being ectothermic, having extended periods of dormancy, having secretive lifestyles, and being dependent on crawling or hopping for transportation. In addition both amphibians and reptiles are often greatly maligned and misunderstood by humans.

An *ectotherm* is a species dependent on the environment to regulate their body temperature. Amphibians and reptiles must actively seek out warmer or cooler areas to maintain a relatively constant body temperature during periods of activity. When weather conditions are unsuitable, they become inactive. Amphibians and reptiles in Minnesota may hibernate up to six months, generally extending from mid-October through mid-April. During this period both amphibians and reptiles substantially reduce their physical activity and cease to feed. One exception to this is Mudpuppies, which actively feed all winter. Metabolic rates slow to conserve stored fat deposits. Some species also enter a period of dormancy called estivation during arid or hot conditions. Dormancy may also occur during periods of water or food shortage, even if temperatures are suitable for activity.

Amphibians and reptiles normally lead secretive lives. Although seasonal breeding and migration activities make them more apparent at certain times of the year, they generally spend much of the time hidden from view in burrows; under logs, leaf litter, or rocks; or at the bottom of ponds. Many amphibians and reptiles are masters of camouflage by remaining motionless, and their cryptic coloration allows them to blend into the surroundings.

Another common denominator of amphibians and reptiles is the restrictions imposed by their size and modes of locomotion. Generally, amphibians and reptiles are small animals that crawl, hop, or swim. Many are capable of surprising bursts of speed, but these peaks of speed are of short duration when compared to birds or mammals. Because amphibians and reptiles are primarily restricted to surface travel, they remain within relatively small territories, and they are highly vulnerable to habitat changes.

Amphibians and reptiles suffer greatly from human ignorance and prejudice. Their secretive nature and their methods of movement provoke fear and hate in many people. Many

The great frog capture. Photograph by Tony Gamble.

The rewards of fieldwork. Photograph by Minnesota Department of Natural Resources—Carol D. Hall.

amphibians and reptiles (especially snakes) are killed needlessly each year by well-meaning but poorly informed people. These animals are an important segment of our biological diversity: they deserve the protection that we provide our more "likable" wildlife. When the public becomes more informed, acceptance and attitudes toward amphibians and reptiles improve.

HISTORY OF HERPETOLOGY IN MINNESOTA

Reptiles and amphibians have been of interest to residents of Minnesota for millennia. Jenks (1936) reported turtle shells from the "Pleistocene man" site, which is estimated to date to approximately 10,000 BP (before the present). While Native

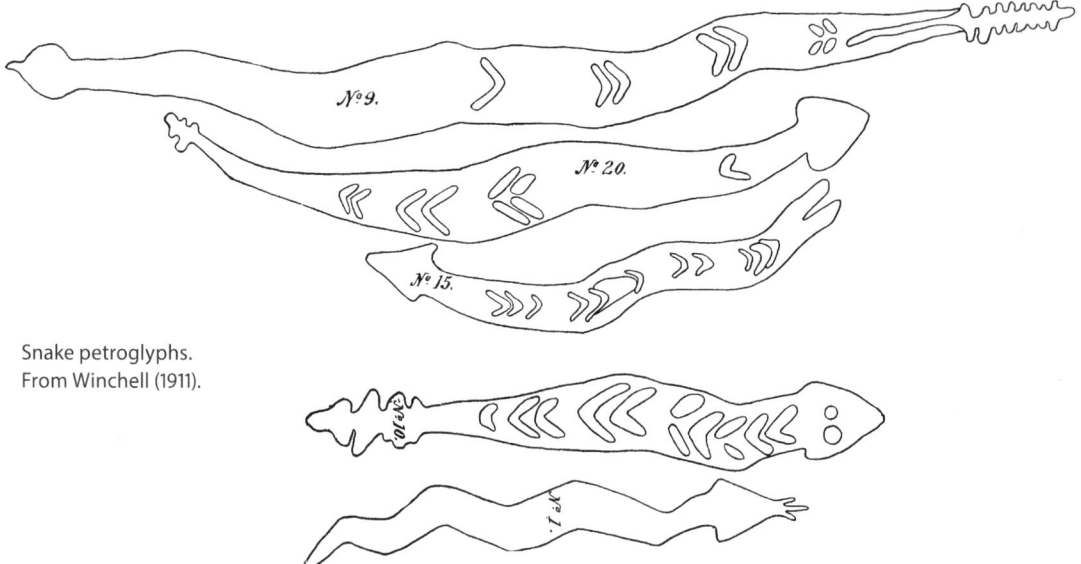

Snake petroglyphs.
From Winchell (1911).

Turtle petroglyphs.
From Winchell (1911).

Americans did not leave a written record, they did leave petroglyphs of turtles and snakes, including rattlesnakes (Winchell 1911). Some of the better petroglyphs are the turtles found at the Jeffers Petrogylphs in Brown County, and the snakes found in the La Moille Cave in Winona, which unfortunately has been destroyed. Petroglyphs at these sites range from 7,000 to 200 BP.

The first reports by Europeans were from Father Hennepin during a French exploration in 1679 and 1680. Hennepin (1698) recorded seeing a huge serpent near St. Anthony Falls, and rattlesnakes near Lake Pepin. Jonathan Carver (1796) traveled through the Midwest, including what would later be Minnesota, and noted seeing snakes and lizards, but he gave no specific identifications or locations.

Reports of explorers in the 1800s include more information on encounters with snakes and turtles, sometimes with erroneous identifications. William H. Keating (1825) reported softshells and map turtles from Lake of the Woods, which is over 200 miles north of their known range. The U.S. Railroad Surveys collected Bullsnakes and Western Foxsnakes from Minnesota in the Fort Snelling area (Baird and Cooper 1859). Many of the specimens were sent to the U.S. National Museum (Smithsonian) and were used in a number of publications by Edward D. Cope (1889, 1900) and others. Very little herpetological work was done in the state between that survey and the early 1930s (Breckenridge 1941).

In the early 1930s, however, several individuals began studies of Minnesota's amphibians and reptiles. The "famous" snake trainer Grace Wiley began her career at the Minneapolis Public Library and Museum. Much of her work was with captive animals, but she was the mentor for several associates who worked on the biology of Minnesota species (Friedrich 1934; Stiles 1938). Gustav Swanson (1935) did the first herpetological study with a statewide perspective. His checklist of amphibians was the foundation for future work on Minnesota amphibians and reptiles.

Walter Breckenridge began his career at the University of Minnesota in the 1930s. His work on Minnesota amphibians and reptiles began as a sideline to his ornithological studies but later expanded into a graduate project on the state's amphibians and reptiles (Breckenridge 1941). Specimens collected for his study became the foundation of the amphibian and reptile collection at the James Ford Bell Museum of Natural History. His graduate project resulted in the publication of *Reptiles and Amphibians of Minnesota* (Breckenridge 1944) and numerous shorter articles (Moriarty and Jones 1988). The book marked the beginning of modern herpetology in Minnesota and has been the standard herpetological reference for Minnesota and the upper Midwest since its publication. Breckenridge's herpetological research continued with studies on softshell turtles (Breckenridge 1960) and toads (Breckenridge and Tester 1961; Tester and Breckenridge 1964; Tester, Parker, and Siniff 1965). The latter work pioneered the use of radioisotope tagging in the study of amphibian movements (Tester 1963, 1964) and produced two theses (Ewert 1969; P. K. Williams 1969).

From the late 1950s to the early 1970s almost all of the published amphibian and reptile work in Minnesota was conducted by graduate students at the University of Minnesota. Studies included the ecology and natural history of Wood Frogs (Bellis 1957; Fishbeck 1968), Green Frogs (Fleming 1976), Mink Frogs (Hedeen 1970), and Prairie Skinks (Nelson 1963). Two important studies were conducted at other institutions. John James (1966) worked on the habits of softshell turtles in the

Early Bullsnake illustration. From U.S. Railroad Survey, 1857.

Walter J. Breckenridge following toads. Photograph from the Bell Museum of Natural History, University of Minnesota.

John Tester measuring toad depths. Photograph from the Bell Museum of Natural History, University of Minnesota.

Mississippi River at Winona State University, and Jeffrey Lang (1971) studied the use of ant mounds by hibernating snakes in northern Minnesota while at North Dakota State University.

Extensive research on Northern Leopard Frogs in Minnesota began during this period. Two researchers, David Merrell and Robert McKinnell, have conducted the majority of the work. Their research, totaling 30 publications (Moriarty and Jones 1988), has focused on the genetics and the biology of cancerous tumors in Northern Leopard Frogs, respectively. Merrell (1977) provided a useful account on the ecology and life history of this species. The interest in frogs expanded greatly with the discovery of deformed frogs in 1995 (Helgen et al. 1998). Numerous studies and theories came out of the deformed frog period (Souder 2000). Deformed frog issues are still a concern (Helgen 2012). Additional information on malformed frogs can be found in the section "Amphibian Declines and Diseases" later in this introduction.

Minnesota Department of Natural Resources

Nongame Wildlife Program

Begun in the late 1970s, the Minnesota Department of Natural Resources (DNR) Nongame Wildlife Program and Natural Heritage Program promoted interest in and research on Minnesota's amphibians and reptiles. Carrol Henderson (1979a, 1979b, 1979c, 1980a, 1980b, 1980c, 1980d, 1980e, 1980f) prepared a series of pamphlets on the distribution of amphibians and reptiles in Minnesota, and the Natural Heritage Program began keeping computer records of the rare species. Together this information was used by the reptile and amphibian group of the Minnesota Endangered Species Technical Advisory Committee, the group responsible for developing the state's first list of endangered, threatened, and special concern amphibians and reptiles (Coffin and Pfannmuller 1988; Lang et al. 1982). The list has been revised several times. The current list (Minnesota DNR 2013) includes 5 amphibians and 11 reptiles. The DNR has also developed the Minnesota Comprehensive Wildlife Conservation Strategy (Minnesota DNR 2006), a plan that covers natural habitats and rare species. The species in this plan are listed as Species of Greatest Conservation Need (SGCN). Six amphibians and 17 reptiles are included as SGCNs. This number is 45 percent of Minnesota's amphibians and reptiles.

The Nongame Wildlife Program has continued to be involved in amphibian and reptile research and management. The program funded studies on Wood Turtles (Ewert 1984, 1985; Buech, Nelson, and Brecke 1990; Oldfield 1988), Blanding's Turtles (Linck 1988; Dorff 1995a; Piepgras 1998; Piepgras et al. 1998; Sajwaj 1998; Lang 2000, 2002; Hamernick 2001), Timber Rattlesnakes (Keyler and Oldfield 1992; Keyler and Fuller 1999; Keyler and Wilzbacher 2002), Gophersnakes (Moriarty 1991),

Five-lined Skinks (Matthews 1990), Northern Leopard Frogs (Hoppe and McKinnell 1989, 1991a, 1991b), and Blanchard's Cricket Frogs (Whitford 1991). A DNR-funded study of the effect of commercial harvest of Painted Turtles (A. B. Gamble 2003; T. Gamble and Simons 2004) was conducted by the University of Minnesota.

Several educational publications were developed to help residents identify and avoid rattlesnakes (Minnesota DNR 1998). The Timber Rattlesnake recovery plan includes ongoing management and a relocation program for rattlesnakes near developed areas. A pamphlet on the snakes and lizards of Minnesota (Perry and Dexter 1989) was also published and revised in 2010 (Christoffel, Edwards, and Perry 2010).

The Minnesota Frog and Toad Calling Survey has become a fixture within the Nongame Wildlife Program. The survey was started in 1993 (Moriarty 1998b) and was moved to the DNR in 1996 with the start of the North American Amphibians Monitoring Program (NAAMP) (Monstad and Baker 2004). The survey is now the largest in the NAAMP system with 220 routes with an average annual completion of 68 percent (150 routes) (H. Cyr, pers. comm.).

Minnesota Biological Survey

Initiated in 1987, the Minnesota County Biological Survey (MCBS) (Natural Heritage Program 1988) has further promoted the study of Minnesota's amphibians and reptiles (Moriarty 1988). Known today as the Minnesota Biological Survey (MBS), it documents the locations of rare amphibians and reptiles, and it is developing monitoring programs for selected reptile populations. MBS biologists collected the first state records for the Four-toed Salamander (Dorff 1995a) and Spotted Salamander (Hall 2002). Since 1987, MBS has contributed nearly 1,000 new records of rare amphibians and reptiles to the DNR's information systems.

Minnesota Herpetological Society

The Minnesota Herpetological Society (MHS) was founded in 1981 following a symposium on the ecology of reptiles and amphibians in Minnesota (Jones 1981; Elwell, Cram, and Johnson

1981). The society has acted as an important source of information on the conservation, natural history, and care of amphibians and reptiles in Minnesota. The Minnesota Nongame Program worked with the MHS on several studies in southeast and southwest Minnesota in the mid-1980s (Moriarty 1985a, 1986). Occasional surveys were held in the 1990s that visited Camp Ripley, Sherburne National Wildlife Refuge, and Kasota Prairie, and in 2001 the MHS reestablished an annual state park survey (LeClere 2011). These field trips have provided valuable information to the DNR's state parks and have exposed numerous MHS members to fieldwork and Minnesota species. The society also published distribution maps of the state's native amphibians and reptiles in the 1980s to stimulate interest in Minnesota species (Moriarty 1985b). The MHS started an occasional paper series in 1985 to publish longer papers of regional interest. The eighth in the series, "A History of MHS Field Surveys" by LeClere, was published in 2011.

The MHS worked with the Nongame Wildlife Program in erecting turtle crossing signs at Weaver Dunes in 1988 to protect Blanding's Turtles and has continued to support habitat protection at Weaver. In 1989, the society led the effort to remove the bounty on rattlesnakes.

In 1995 the MHS convened a conference on the conservation and status of Minnesota's amphibians and reptiles (Moriarty and Jones 1997). This conference brought together 17 speakers presenting research on a variety of native species.

The MHS was a sponsor of the Blanding's Turtle Workshop held at the Bell Museum of Natural History (Moriarty 1998a, 2000), attended by turtle biologists from 10 states and Canada, which highlighted the extensive Blanding's Turtle research being conducted in Minnesota. The society was also a major sponsor of the 2005 symposium on timber rattlesnake biology and conservation in the upper Mississippi River valley (Keyler and Cochran 2005). Daniel Keyler's (2005) MHS occasional paper on snake bites in the upper Midwest was released at the symposium. The MHS has hosted the Midwest Herpetological Symposium five times since the 1980s. This symposium is a mixture of conservation and herpetoculture.

The MHS has partnered with the Nongame Wildlife Program to provide a reptile display at the Minnesota State Fair. It is one of the most popular exhibits in the DNR Building.

Other Herpetological Efforts

The study of herpetology has been particularly active in Minnesota since 1980. Between 1900 to 1980, 244 articles were published on Minnesota herpetology (Moriarty and Jones 1988). Comparatively, there have been over 800 citations since 1980 (Moriarty 2004, 2007). This proliferation is due to an increase in research and fieldwork at the DNR, University of Minnesota, and other institutions. The MHS newsletter has also generated a number of citations.

The University of Minnesota's Bell Museum of Natural History has more than 15,000 specimens in the amphibian and reptile collection. Ken Kozak, curator of herpetology at the Bell Museum, has a number of students looking into amphibian conservation issues in Minnesota and other states. Mark Bee is studying the communication strategies of frogs in Minnesota.

In the past decade several other Minnesota universities have hired herpetologists. Phil Cochran at St. Mary's University at Winona is studying rattlesnakes and other local species. Steve Freedberg at St. Olaf College is working on turtles in the Mississippi River. Tim Lewis at the University of St. Thomas is studying urban turtle populations, and Donna Stockrahm at Minnesota State University Moorhead is working on Painted Turtles in northwestern Minnesota.

AMPHIBIAN AND REPTILE HABITATS

Amphibians and reptiles depend on a diversity of ecologically healthy habitats. Information about these habitats combined with knowledge of the geographic distribution of species informs herpetologists about where they can expect to find amphibians and reptiles. For example, knowledge about Common Five-lined Skinks tells herpetologists that they will likely find these skinks by exploring granite or limestone outcrops in southeast Minnesota, whereas searching for this species in coniferous forests in northern Minnesota would be a waste of time.

A hierarchy of ecological classification has been developed for Minnesota that is useful for predicting where a species may occur in the state. The Ecological Classification System (ECS) is separated into provinces, sections, and subsections. The provinces, the highest level, are defined by major climate zones, native vegetation, and biomes. The 4 provinces are Prairie Parkland, Tallgrass Aspen Parkland, Eastern Broadleaf (Deciduous) Forest, and Laurentian Mixed (Coniferous) Forest. These provinces are divided into 10 sections based on glacial deposits, regional elevations, major vegetation patterns, and regional climate. These sections are then divided into 26 subsections based on more detailed glacial and geologic characteristics and native plant communities. Amphibians and reptiles tend to be found in multiple ecological sections (Table A) because their habitats cross section boundaries. The map of the natural vegetation of Minnesota, which shows the original extent of habitats prior to European settlement (Wendt and Coffin 1988), illustrates that plant communities cross biomes, which include both terrestrial and aquatic habitats.

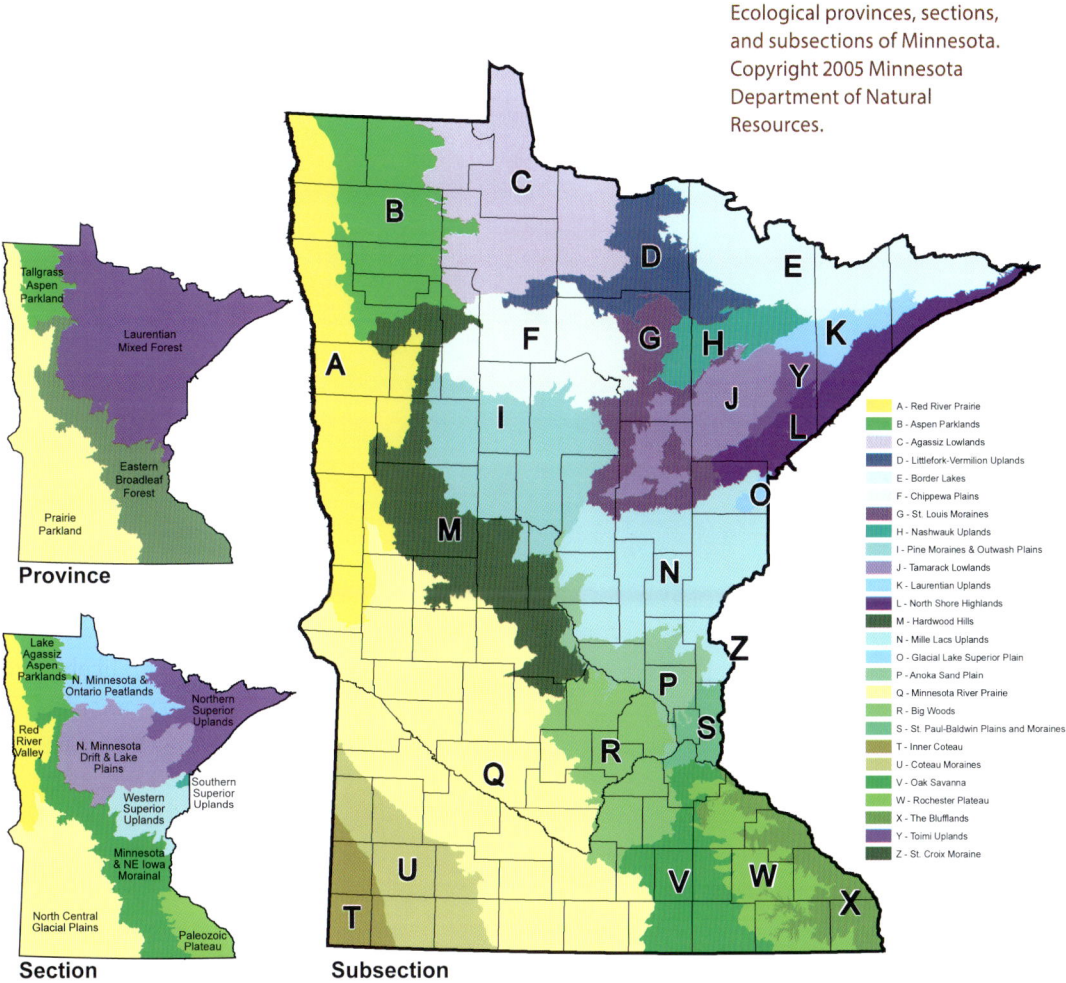

Ecological provinces, sections, and subsections of Minnesota. Copyright 2005 Minnesota Department of Natural Resources.

Province

Section

Subsection

A - Red River Prairie
B - Aspen Parklands
C - Agassiz Lowlands
D - Littlefork-Vermilion Uplands
E - Border Lakes
F - Chippewa Plains
G - St. Louis Moraines
H - Nashwauk Uplands
I - Pine Moraines & Outwash Plains
J - Tamarack Lowlands
K - Laurentian Uplands
L - North Shore Highlands
M - Hardwood Hills
N - Mille Lacs Uplands
O - Glacial Lake Superior Plain
P - Anoka Sand Plain
Q - Minnesota River Prairie
R - Big Woods
S - St. Paul-Baldwin Plains and Moraines
T - Inner Coteau
U - Coteau Moraines
V - Oak Savanna
W - Rochester Plateau
X - The Blufflands
Y - Toimi Uplands
Z - St. Croix Moraine

Climate also affects the distribution and abundance of amphibians and reptiles. Minnesota is colder to the north and drier to the west. The species richness of amphibians and reptiles, especially the latter, decreases to the north and west in Minnesota. An extreme example is the species richness of amphibians and reptiles in the southeast (Houston County, 12 amphibians and 22 reptiles) compared to the species richness in the northwest (Kittson County, 10 amphibians and 7 reptiles). Arid conditions

Natural vegetation of Minnesota. Adapted from Marschner (1974).

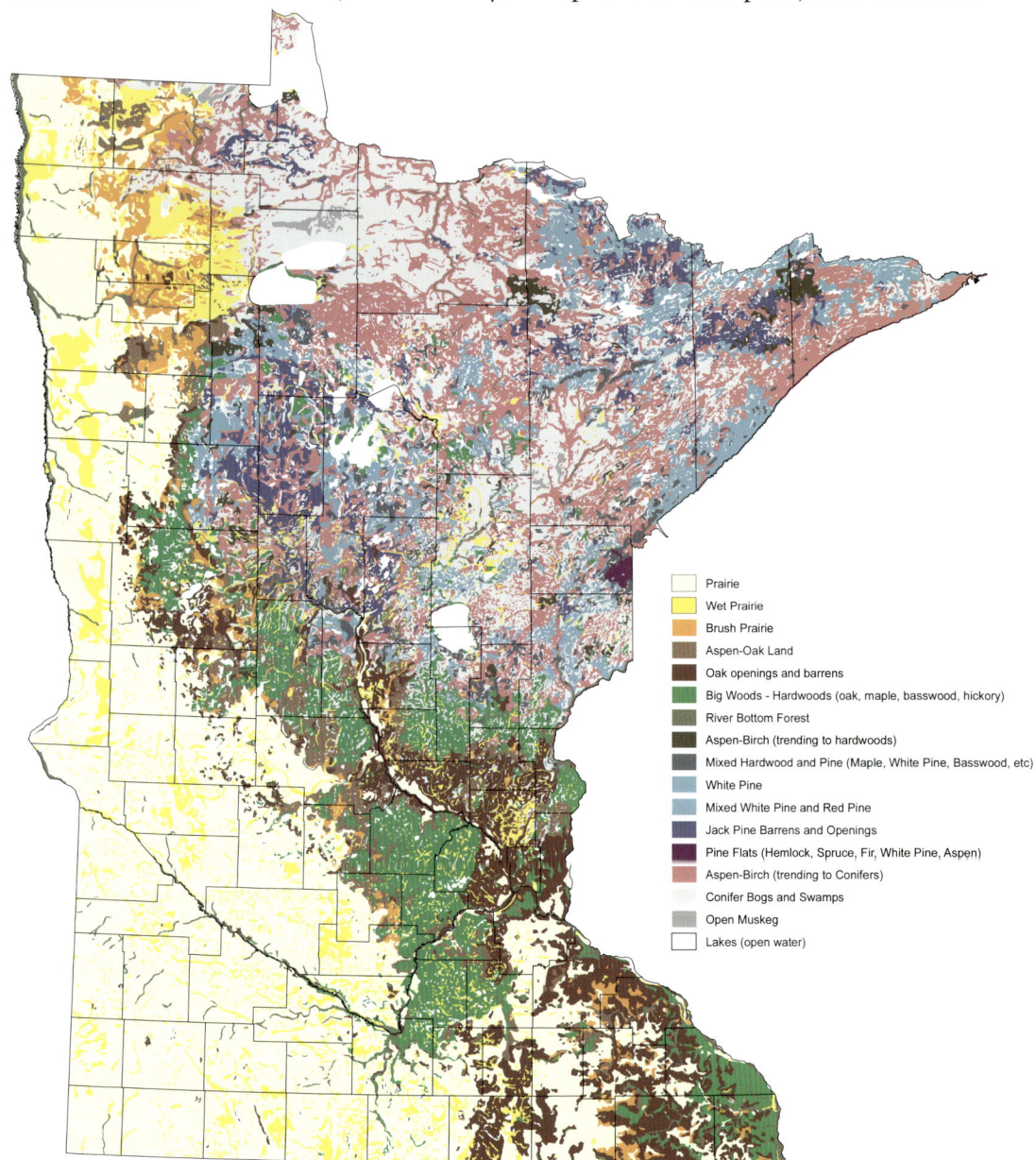

- Prairie
- Wet Prairie
- Brush Prairie
- Aspen-Oak Land
- Oak openings and barrens
- Big Woods - Hardwoods (oak, maple, basswood, hickory)
- River Bottom Forest
- Aspen-Birch (trending to hardwoods)
- Mixed Hardwood and Pine (Maple, White Pine, Basswood, etc)
- White Pine
- Mixed White Pine and Red Pine
- Jack Pine Barrens and Openings
- Pine Flats (Hemlock, Spruce, Fir, White Pine, Aspen)
- Aspen-Birch (trending to Conifers)
- Conifer Bogs and Swamps
- Open Muskeg
- Lakes (open water)

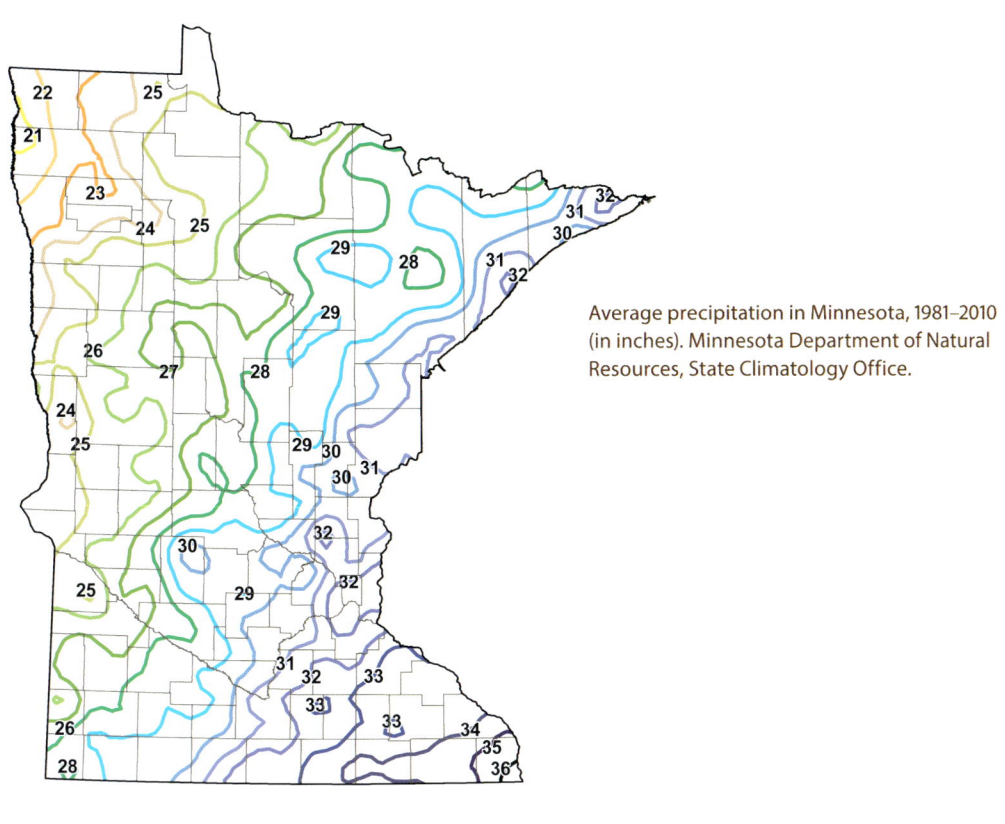

Average precipitation in Minnesota, 1981–2010 (in inches). Minnesota Department of Natural Resources, State Climatology Office.

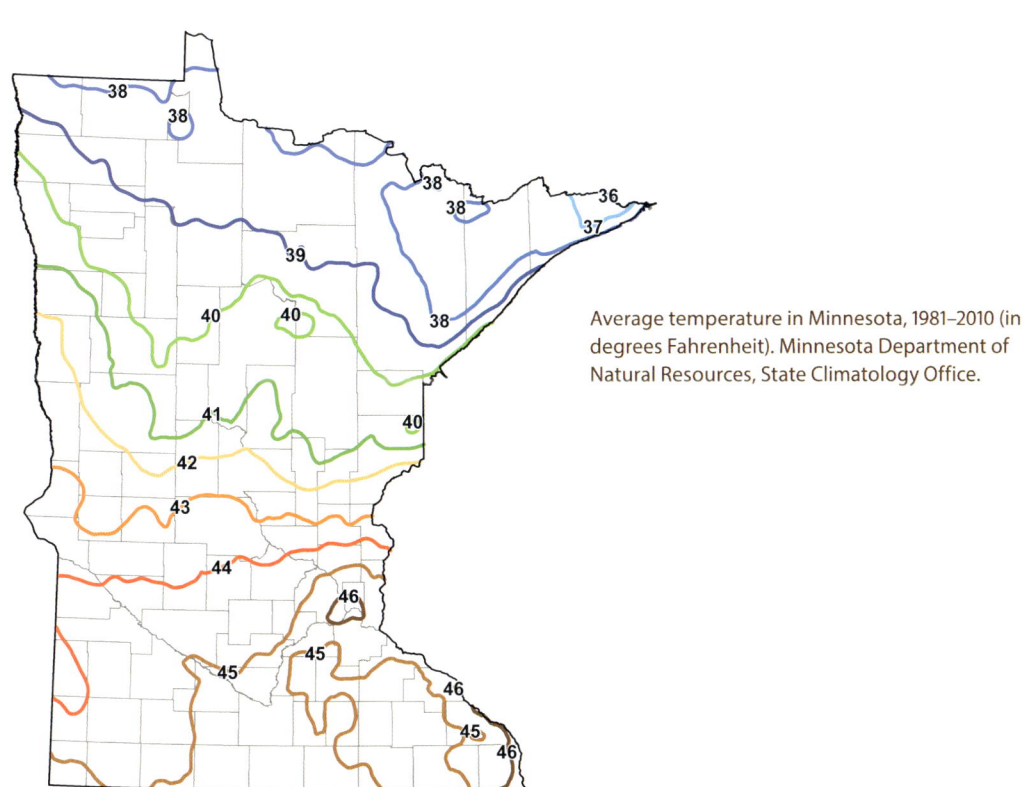

Average temperature in Minnesota, 1981–2010 (in degrees Fahrenheit). Minnesota Department of Natural Resources, State Climatology Office.

Table A. Amphibian and reptile distributions by ecological section

	Prairie Parkland		Tallgrass Aspen Parkland	Eastern Broadleaf Forest		Laurentian Mixed Forest				
AMPHIBIANS	North Central Glaciated Plains	Red River Valley	Lake Agassiz Aspen Parklands	Paleozoic Plateau	Minnesota & NE Iowa Morainal	Southern Superior Uplands	Western Superior Uplands	Northern Superior Uplands	Northern Minnesota & Ontario Peatlands	Northern Minnesota Drift & Lake Plains
American Toad	X	X	X	X	X	X	X	X	X	X
Great Plains Toad	X	X	X							
Canadian Toad	X	X	X							
Blanchard's Cricket Frog	X			X	X					
Cope's Gray Treefrog	X		X	X	X					X
Gray Treefrog	X	X	X	X	X		X	X	X	X
Spring Peeper			X	X	X	X	X	X	X	X
Boreal Chorus Frog	X	X	X	X	X	X	X	X	X	X
American Bullfrog	X			X	X		X			
Green Frog				X	X	X	X	X	X	X
Pickerel Frog				X						
Northern Leopard Frog	X	X	X	X	X	X	X	X	X	X
Mink Frog			X		X	X	X	X	X	X
Wood Frog	X	X	X	X	X	X	X	X	X	X
Blue-spotted Salamander	X			X	X	X	X	X	X	X
Spotted Salamander							X			
Western Tiger Salamander	X									
Eastern Tiger Salamander	X	X	X		X		X			X
Four-toed Salamander							X			X
Eastern Red-backed Salamander	?					X	X	X		X
Mudpuppy	X	X	X	X	X		X			X
Eastern Newt	X			X	X	X	X	X	X	X
Total	**16**	**9**	**12**	**14**	**15**	**10**	**16**	**11**	**10**	**15**
REPTILES										
Common Five-lined Skink	X			X	X		X			
Prairie Skink	X	X	X	X	X		X			X
Six-lined Racerunner				X	X					
North American Racer	X			X	X					

Species										
Ring-necked Snake				X			X			
Plains Hog-nosed Snake	X	X		X	X		X			X
Eastern Hog-nosed Snake				X	X		X		X	X
Milksnake	X			X	X		X			
Common Watersnake				X	X		X			
Smooth Greensnake	X	X	X	X	X		X	X		X
Western Ratsnake				X			X			
Western Foxsnake	X			X	X		X			X
Gophersnake	X	X	X	X	X		X			X
Dekay's Brownsnake	X	X		X	X		X	X		X
Red-bellied Snake	X	X	X	X	X	X	X	X	X	X
Plains Gartersnake	X	X	X	X	X	X	X	X		X
Common Gartersnake	X	X	X	X	X	X	X	X	X	X
Lined Snake	X									
Timber Rattlesnake				X	X					
Massasauga				X						
Snapping Turtle	X	X		X	X	X	X	X		X
Painted Turtle	X	X	X	X	X	X	X	X	X	X
Blanding's Turtle	X			X	X	X	X			X
Wood Turtle			X	X	X	X	X	X		X
Northern Map Turtle	X			X	X		X			
Southern Map Turtle				X	X					
False Map Turtle	X			X	X					
Pond Slider				X						
Eastern Musk Turtle				X						
Smooth Softshell	X			X	X		X			X
Spiny Softshell	X			X	X		X			X
Total	**20**	**9**	**8**	**30**	**25**	**7**	**16**	**6**	**3**	**12**
Amphibian total	16	9	12	14	15	10	16	11	10	15
Reptile total	20	9	8	30	25	7	16	6	3	12
Total	**36**	**18**	**20**	**44**	**40**	**17**	**32**	**17**	**13**	**27**

and the limited availability of habitat that can provide adequate frost-free overwintering sites in the northwest are likely responsible for the smaller number of species.

Amphibian and reptile habitats in Minnesota can be divided into eight types, four aquatic and four terrestrial, which cover the spectrum of natural vegetation in the state. Aquatic habitats include rivers and streams, lakes and ponds, marshes and prairie wetlands, and peatlands. Terrestrial habitats include three forest types and one grassland. Most of Minnesota's amphibians and turtles live in aquatic habitats, whereas lizards and the majority of snakes occupy terrestrial habitats. Each habitat has its own diversity of amphibians and reptiles (Table B). Within each habitat type amphibians and reptiles are not equally distributed. Food, cover, the availability of breeding sites, and annual climate variation affect their distribution and abundance.

Watersheds

Watershed boundaries are defined by the flow of water across the land into a single aquatic unit. In Minnesota three Continental Divides separate surface water runoff into three major drainage basins: Hudson Bay, the Great Lakes, and the Mississippi River. These major basins are subdivided into 9 river basins and 81 subbasins, or watersheds. The health of each watershed is based on the patterns of water flow, the geology and landscape, plant and animal communities, land use, and risk factors such as contaminants. The abundance and health of wetlands, lakes, and rivers are critical to the survival of our amphibians and reptiles. Except for the Mudpuppy, the distribution of these species is rarely limited by watershed boundaries, for river and stream corridors provide connectivity between plant communities and important features such as breeding, nesting, and foraging areas. In doing so, they greatly influence the dispersal of these species across the landscape.

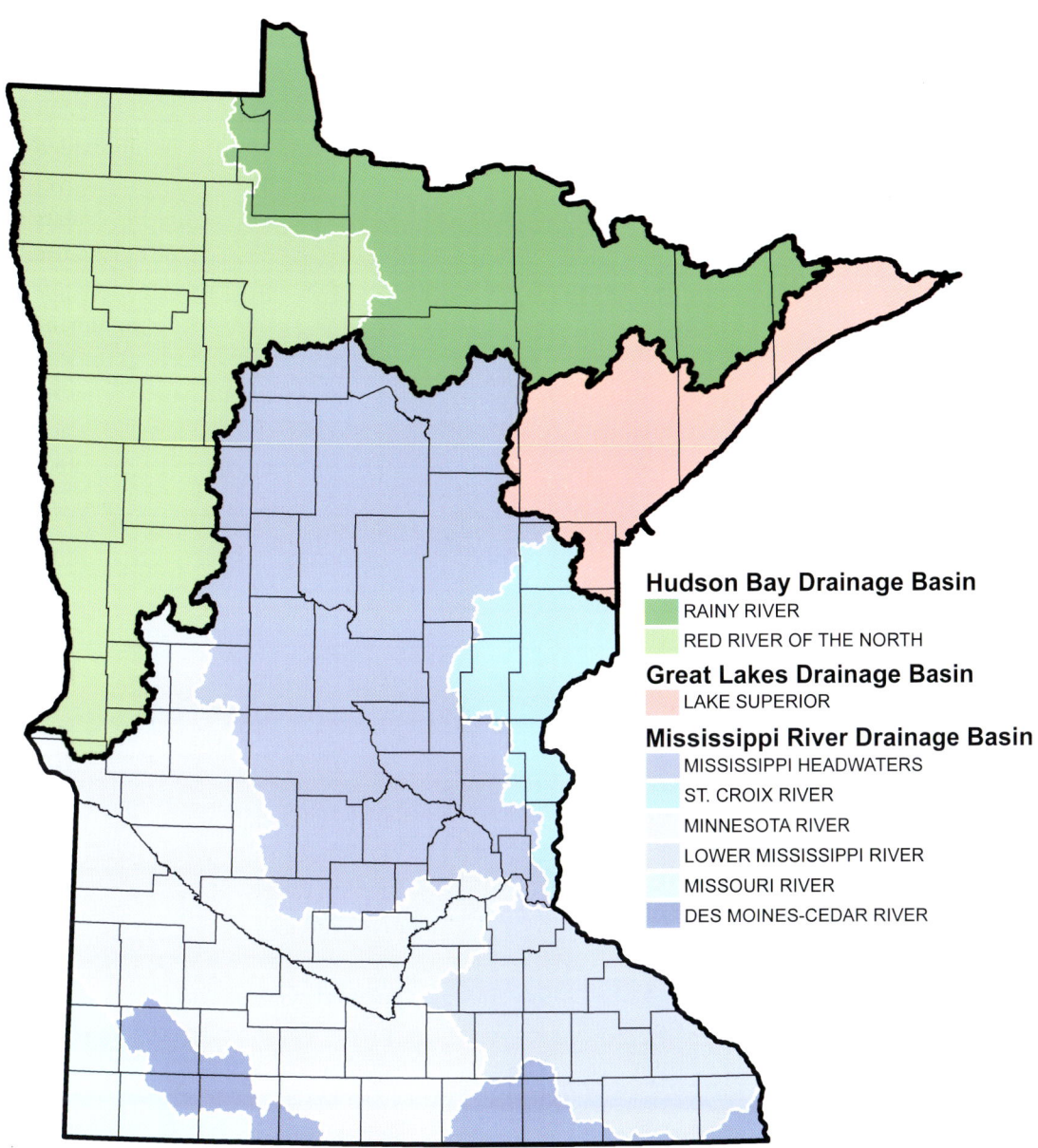

Hudson Bay Drainage Basin
  ■ RAINY RIVER
  ■ RED RIVER OF THE NORTH
Great Lakes Drainage Basin
  ■ LAKE SUPERIOR
Mississippi River Drainage Basin
  ■ MISSISSIPPI HEADWATERS
  ■ ST. CROIX RIVER
  ■ MINNESOTA RIVER
  ■ LOWER MISSISSIPPI RIVER
  ■ MISSOURI RIVER
  ■ DES MOINES-CEDAR RIVER

Major watersheds of Minnesota.

Table B. Amphibian and reptile distributions by habitat type

	Aquatic					Terrestrial				
AMPHIBIANS	Rivers and streams	Lakes and ponds	Marshes and prairie wetlands	Peatlands	Floodplain forest	Big Woods and oak woods	Coniferous–N. hardwood forest	Prairie	Agricultural	Developed
American Toad	X	X	X	X	X	X	X	X	X	X
Great Plains Toad		X	X					X		
Canadian Toad	X	X	X		X			X	X	
Blanchard's Cricket Frog	X	X								
Cope's Gray Treefrog	X	X	X		X	X				X
Gray Treefrog	X	X	X	X	X	X	X			X
Spring Peeper		X	X	X	X	X	X			
Boreal Chorus Frog		X	X	X		X	X	X	X	X
American Bullfrog	X									
Green Frog	X	X	X	X						
Pickerel Frog	X	X								
Northern Leopard Frog	X	X	X	X	X	X	X	X	X	
Mink Frog	X	X								
Wood Frog		X	X	X	X	X	X			
Blue-spotted Salamander		X	X	X	X	X	X			
Spotted Salamander		X	X				X			
Western Tiger Salamander		X	X					X	X	
Eastern Tiger Salamander		X	X		X	X		X	X	X
Four-toed Salamander				X			X			
Eastern Red-backed Salamander							X			
Mudpuppy	X	X								
Eastern Newt	X	X	X		X		X			
Total	**12**	**19**	**15**	**8**	**10**	**8**	**11**	**8**	**6**	**5**
REPTILES										
Common Five-lined Skink						X		X		
Prairie Skink						X		X		X
Six-lined Racerunner								X		
North American Racer						X		X		

Species	1	2	3	4	5	6	7	8	9
Ring-necked Snake				X				X	
Plains Hog-nosed Snake				X				X	
Eastern Hog-nosed Snake					X			X	
Milksnake				X	X			X	X
Common Watersnake	X		X		X				
Smooth Greensnake		X			X			X	
Western Ratsnake					X				
Western Foxsnake					X				
Gophersnake								X	X
Dekay's Brownsnake				X	X			X	X
Red-bellied Snake			X		X	X		X	X
Plains Gartersnake	X		X		X	X	X	X	X
Common Gartersnake			X		X			X	X
Lined Snake								X	
Timber Rattlesnake					X			X	
Massasauga	X	X			X				
Snapping Turtle	X	X							X
Painted Turtle	X	X	X					X	X
Blanding's Turtle	X	X						X	X
Wood Turtle	X				X	X			
Northern Map Turtle	X								
Southern Map Turtle	X								
False Map Turtle	X								
Pond Slider	X	X						X	X
Eastern Musk Turtle	X								
Smooth Softshell	X								
Spiny Softshell	X								
Total	**13**	**6**	**7**	**3**	**11**	**14**	**3**	**17**	**8**
Amphibian total	12	19	15	8	10	8	11	8	5
Reptile total	13	6	7	3	11	14	3	17	8
Total	**25**	**25**	**22**	**11**	**21**	**22**	**14**	**25**	**13**

Aquatic Habitats

Rivers and Streams

Minnesota has a large network of streams and rivers within the drainage basins of the Minnesota, Mississippi, and Red Rivers and Lake Superior. Rivers and streams have various physical characteristics that are habitat for specific species. For example, a substrate of bedrock and gravel, such as in trout streams, provides habitat for Wood Turtles and Mudpuppies. Rivers with thick sand and organic bottoms, such as the lower Mississippi River, are the major habitat for map turtles and softshells.

Rivers and streams may have a network of backwaters, oxbows, and coves. These areas tend to have limited current and dense vegetation, providing important habitat for a variety of amphibians and reptiles. The backwaters of the Mississippi River are the only habitat in Minnesota where American Bullfrogs naturally occur.

Small river with wooded riparian zone. Photograph by Tony Gamble.

Floodplain forest. Photograph by Minnesota Department of Natural Resources—Michael D. Lee.

Lakes and Ponds

Lakes and ponds are bodies of water that range in size from Lake Superior to ponds less than 1 hectare (2.5 acres). Smaller lakes and shallow bays of large lakes provide better habitat for amphibians and reptiles than does deep, open water. The shallow water supports more vegetation, both emergent and submergent, which provides cover and food. Warmer water temperatures in shallow areas also promote the development and growth of amphibian larvae. Small ponds lack predatory fish and provide good habitat for tadpoles and salamander larvae.

Marshes and Prairie Wetlands

Marshes are large areas of shallow water with thick emergent vegetation, mainly cattails, sedges, and bulrushes. These areas are commonly associated with prairie lowlands in the western half of the state. Cattail marshes are scattered throughout the state in

Beaver pond. Photograph by Minnesota Department of Natural Resources—Barb Delaney.

Shallow lake with water lilies. Photograph by Minnesota Department of Natural Resources—Carol D. Hall.

association with lake systems. Prairie wetlands are important breeding sites for Great Plains and Canadian Toads. Marshes provide good amphibian and reptile habitat for a variety of species because dense vegetation provides cover, water, and food. Many of these wetlands have been lost to drainage for agriculture.

Peatlands

Peatlands can be divided into bogs and fens. The sphagnum beds of bogs are normally water saturated and sometimes contain floating mats of vegetation. Bog water is very acidic (pH < 4.5) and nutrient poor, characteristics that may limit the number of amphibians that can use the habitat. Wood Frogs, Mink Frogs, American Toads, and Blue-spotted Salamanders are species commonly found in peatlands.

Calcareous fens are scattered across the state but are very rare. Blanding's Turtles are found in association with these calcareous habitats in southwestern Minnesota.

Sedge meadow. Photograph by Minnesota Department of Natural Resources—Carol D. Hall.

Drying wetlands. Photograph by Minnesota Department of Natural Resources—Carol D. Hall.

Peatland. Photograph by Minnesota Department of Natural Resources—Kurt Rusterholz.

Terrestrial Habitats

Floodplain Forest

Floodplain or bottomland forests are associated with the floodplains of large rivers. Their moist rich soils, which may be seasonally flooded, promote a lush growth of vegetation. Major tree species associated with this community are elms, maples, ash, and cottonwood. Willows are normally found along the riverbanks. Wood Turtles are found in bottomland forests that border streams where raspberries and strawberries, favorite foods of the turtle, are common. Red-bellied Snakes and Dekay's Brownsnakes are also found in this forest type because of high soil invertebrate populations.

Bottomland forest. Photograph by Minnesota Department of Natural Resources—Barb Delaney.

Big Woods and Oak Forest

Big woods and oak forests are the two main deciduous forest types found in central and southeastern Minnesota. These forests vary in soil conditions and topography. The southeastern region of the state, where oak *(Quercus)* forests dominate, lies in the "driftless zone," an unglaciated area of steep hills and rock outcrops. These outcrops are an important habitat component for many southeastern species, including the Timber Rattlesnake, Common Five-lined Skink, and Western Ratsnake.

The big woods are on a glacial plain with rolling low hills dominated by elms, maples, and basswoods. Woodland species of amphibians and reptiles, including American Toads, Wood Frogs, Common Gartersnakes, and Red-bellied Snakes, are dependent on logs and downed branches in these forests for cover. Many forest stands, however, have been degraded for amphibians and reptiles because all of the undergrowth and downed logs have been removed for grazing or firewood. The increase of invasive species, such as buckthorn, garlic mustard, and earthworms, has also degraded this and other habitats.

Northern hardwoods forest. Photograph by Minnesota Department of Natural Resources—Carol D. Hall.

Forest floor with woody debris. Photograph by Minnesota Department of Natural Resources—Scott Zager.

Coniferous forest floor.
Photograph by Minnesota
Department of Natural
Resources—Kurt Rusterholz.

Coniferous–Northern Hardwood Forest

Conifer forests, including red and white pine, jack pine, and spruce-fir, are found in northern Minnesota. Dominated in pre-settlement time by spruce, fir, and pine, large areas of this forest type have been converted to aspen and birch hardwood stands through commercial logging. Although conifer forests are generally poor habitat for amphibians and reptiles, Red-backed Salamanders and Blue-spotted Salamanders are regularly found in hardwood forest stands resting under logs and pieces of bark.

Prairies

Prior to European settlement, western Minnesota was dominated by prairie and vast grasslands with a few scattered trees near streams and lakes. Found on various soils and substrates, prairies are dominated by bluestem, Indian grass, and switchgrass but also have a wide diversity of forbs (wildflowers). Today 99 percent of Minnesota's original prairie grasslands have been

Beach ridge prairie.
Photograph by Minnesota
Department of Natural
Resources—Tim Whitfield.

converted to agricultural use. There are many open grasslands throughout Minnesota that are not prairies because they have been planted with nonnative grass species. Amphibians and reptiles that are restricted to prairies, including the Plains Hognosed Snake, Gophersnake, and Canadian Toad, tend not to use the nonnative grasslands (J. Tester, pers. comm.). Bluff prairies, a special type of prairie that occurs on steep rocky slopes and rock outcrops in the southeastern part of the state, are the preferred habitat of North American Racers, Ring-necked Snakes, Timber Rattlesnakes, Milksnakes, and Six-lined Racerunners.

Mesic prairie. Photograph by
Minnesota Department of
Natural Resources.

Prescribed burn on a prairie. Photograph by Minnesota Department of Natural Resources—Kelly Lynch Pharis.

Disturbed Habitats

Agricultural Lands

Agricultural habitats include tilled croplands and pastured grasslands. Tilled croplands provide little to no habitat for amphibians and reptiles. Amphibians will breed in flooded farm fields but are rarely successful. Extensive ditching and tiling (underground drainage systems) remove the water from the fields quickly, trapping any tadpoles that may have hatched. Turtles will nest in agricultural fields, but these nests tend to be less successful (Freedberg et al. 2011).

Pastured grasslands can provide habitat for many species as long as the area is not overgrazed or drained of wetlands. Stock ponds can be very good habitat for salamander breeding. Concentrated cattle use tramples the soil, making it too compact for amphibians and reptiles to burrow in.

Agricultural fields in western Minnesota. Photograph by Minnesota Department of Natural Resources—Gerda Nordquist.

Livestock negatively impact stream habitats. Photograph by Minnesota Department of Natural Resources—Carol D. Hall.

Urban and Suburban Habitats

Urban and suburban areas are where most Minnesotans live. Some amphibians and reptiles can survive in these developed habitats. These include American Toads, Boreal Chorus Frogs, Painted Turtles, Snapping Turtles, Prairie Skinks, and Red-bellied Snakes. The downtown portions of St. Paul and Minneapolis do not provide any habitat, but larger parks within both cities have relatively diverse populations of amphibians and reptiles. Species needing wooded wetlands, such as Spring Peepers and Wood Frogs, do not fare well. Neither do large snakes, like Western Foxsnakes and Gophersnakes. Also not common in these habitats are those species with annual breeding migrations, which include Northern Leopard Frogs, Eastern Tiger Salamanders, and Blanding's Turtles. These species have increased mortality on roads and have their movements blocked by buildings, walls, and roads.

Gartersnake trapped in erosion netting. Photograph by Minnesota Department of Natural Resources—Carol D. Hall.

Developed areas have hazards, even at sites being restored to natural habitats. Many landscaping and revegetation projects use erosion mats that can contain plastic mesh. Snakes and other wildlife become entangled in the mesh and die. Erosion mats containing this mesh should not be used, and many resource agencies have developed policies prohibiting its use.

OBSERVING AND STUDYING AMPHIBIANS AND REPTILES

Ethical Field Methods

How does a person find amphibians and reptiles in their indigenous habitats? Turtles are readily spotted at a distance basking on logs in ponds or along riverbanks. Frogs are often seen at the edges of ponds leaping into the water or are heard calling from a wetland. Nearly every child at one time or another has caught a toad near their doorstep, in a garden, or at a local park. Nevertheless, many Minnesota species are secretive and not easily found. Even people who spend a great amount of time outdoors rarely see snakes, lizards, or salamanders. The only snake most Minnesotans ever see roaming free is a gartersnake. Lizards and salamanders are secretive and remain hidden much of the time. How does a person go about finding and studying these interesting animals?

Successful herping requires patience, diligence, and time. Obviously, selecting suitable areas to search is the first consideration. Abandoned farmsteads, old rock quarries, woodlots, ponds, and marshes are good spots to investigate. Good sites for looking for amphibians and reptiles may be on private property, and permission from the owner needs to be acquired before beginning the hunt.

Frog search. Photograph by Tony Gamble.

Walk slowly through promising locales, and keep an eye out for movement. Carefully pull aside rocks, logs, shingles, tin, and boards to see what is underneath. Hay hooks and potato rakes are helpful for turning this material over. Pull ground cover toward you to provide a shield in case a wasps' nest, skunk, or a venomous snake is uncovered. The likelihood of uncovering a venomous species in Minnesota is quite rare, but stinging insects are not scarce. Carefully look at the undersurface of the log or board as well as the exposed ground. Small animals often remain still and are difficult to see. Replace cover objects in their original location and position to preserve the microhabitat of the organisms using this living space. Searching for amphibians and reptiles is destructive if not conducted in an environmentally conscientious manner. Treat the habitat kindly!

Swamps, marshes, streams, and sloughs are excellent areas to search. Amphibians and reptiles that are likely to be found in and near wet areas include turtles, watersnakes, frogs, and amphibian larvae. Waders or hip boots give additional protection in cool weather; shorts and old tennis shoes work well in warm weather.

For those seeking adventure, the investigation of breeding congregations of frogs or toads is an exciting and rewarding experience. A good flashlight or headlamp is needed for nighttime searches. Calling amphibians pay little or no attention to a flashlight, so it is easy to observe their calling and breeding behavior. If they stop calling, wait until the vocalizations begin again before proceeding with the search. Toads display little fear during courtship activities, and in no time the males will be calling unabated within hand's reach. Treefrogs are small and have concealing color patterns and thus require more diligence to locate. Scan the area slowly with a flashlight, and look for a reflection from their inflated vocal sacs.

Road driving is another useful technique for finding amphibians and reptiles, particularly at night. During spring and fall migrations, road driving can provide success in Minnesota, but at other times of the year very few amphibians and reptiles are found on roads. Rain events often trigger movements of amphibians and reptiles and can greatly increase chances of success. Choose a rural blacktop road with little traffic. Drive slowly, and keep an eye on the road for small animals. Watch out for other traffic when making a hasty stop. Unfortunately, many of the amphibians and reptiles found on roads are dead. One quickly gains an appreciation for the large numbers of animals that are run over. Roadkills may be of scientific value, especially to establish locality records. Refer to the maps in the species accounts for the currently known distribution. Road-killed specimens from new localities or remains of rare species should be frozen and delivered to the Bell Museum of Natural History. Include an accurate location description and map with the specimen.

Proper timing contributes to the success of a herping trip. Spring is the best season to find amphibians and reptiles, as they emerge from overwintering and subsequently breed. Summer is less rewarding because of decreased activity levels and estivation. Also, heavy vegetation growth provides cover for the animals, making them more difficult to locate. One notable exception during midsummer is the movement of large numbers of newly metamorphed amphibians away from breeding ponds. Road driving on warm summer evenings may be helpful in finding nocturnal species. Many species become more conspicuous in the fall as they migrate to hibernating sites.

Amphibians and reptiles start becoming active by midmorning during warmer months. Air temperatures from 21° to 32°C (70° to 90°F) are preferable, especially for reptiles. Amphibians are active at somewhat cooler temperatures and during rainy weather.

Successful searching requires experience and patience, but luck is a key factor. Even a seasoned veteran has days of failure. An excellent publication describing detailed herpetological techniques is *Field Herpetology: Methods for the Study of Amphibians and Reptiles in Minnesota* (Karns 1986).

Frequently it is necessary to have an animal in hand to make an identification. Care should be taken to prevent injury to the animal during capture and handling. Small amphibians and reptiles may accidentally be killed by overturning rocks, and big hands can unintentionally crush small, wiggly salamanders and lizards. Small amphibians and lizards can be held for examination by gently pressing one of their rear legs between thumb and forefinger while the animal remains free to cling to the remainder of the hand. Large frogs should be grasped around their abdomens with their rear legs extended. Wetting one's hands prior to catching amphibians prevents the loss of protective coating (slime) on frogs and salamanders.

Large nonvenomous snakes should be held with both hands

to keep the frightened animal from injuring itself by thrashing. One hand is used to keep the head under control while the other supports the snake's body. A Snapping Turtle should be held by its rear legs or the back edge of its shell with the head of the turtle pointed away from your body. Suspending a Snapping Turtle solely by its tail may permanently damage the turtle's spinal cord. Softshell turtles can be restrained by grasping the rear edge of their upper shell or in front of the hind legs while staying clear of their sharp beak and claws. Conant and Collins (1998) and Karns (1986) illustrate proper methods of handling amphibians and reptiles.

After the animal is identified, observed, and photographed, it should be released at the capture site. Many amphibians and reptiles are territorial with specific home ranges and die or do poorly in a different locality. Nonnative species and long-term captives should not be released; in most cases these animals quickly perish. On rare occasions, an exotic species may become established in an area and could be detrimental to the native fauna. For example, Pond Sliders have been intentionally released by pet owners in many states and have become established in numerous areas, including Minnesota. In some states they have displaced native turtles.

The best method of transporting snakes and lizards is a well-sewn cotton sack, such as a pillow case. Tie a snug overhand knot in the neck of the sack to prevent escape. Amphibians are best placed in small bottles or self-sealing bags with damp moss or vegetation to prevent desiccation. Amphibians and reptiles trapped in containers in the direct sun perish quickly from overheating. Extreme care must be taken to prevent such unfortunate accidents. Place sacks and containers with occupants in a picnic cooler placed in the shade. These coolers make ideal transportation boxes as they provide protection from temperature extremes and substantially decrease the possibility of an escape.

Species listed as threatened or endangered in Minnesota have full legal protection. Permits are required to study these species. They can be photographed and observed but not collected or harmed. All amphibians and reptiles are protected in national parks, state parks, and state Scientific and Natural Areas. There are restrictions on and license requirements for capturing frogs to be used for purposes other than fishing bait. Possession and collection of turtles for food are regulated.

Contact the DNR for current rules and regulations concerning Minnesota's amphibians and reptiles.

Searching for amphibians and reptiles can be a rewarding and interesting venture. Genuine curiosity about the natural world and a good deal of patience are the key ingredients for countless hours of enjoyment. Information gathered through searches can be an important contribution to conservation efforts.

Field Study

There is a tremendous lack of information concerning the secretive lives of amphibians and reptiles. Information is needed on population dynamics, habitat requirements, predator-prey relationships, reproduction, and ecology. Without sufficient knowledge about the natural history of a species, resource managers have difficulty implementing beneficial short- and long-term conservation programs.

Before going out in the field, take time to study background materials on species of interest. Such preparation saves time and makes observations more meaningful. The requirements for

Salamander searches. Photograph by Minnesota Department of Natural Resources—Kelly Lynch Pharis.

herpetological field study are as simple as curiosity, a notebook, and good references. In addition, binoculars are useful for observing difficult-to-approach animals such as turtles and lizards. Record the locality, weather information, and a description of the habitat when an amphibian or reptile is found. Take notes on size and sex of the animal, unusual markings, and its behavior. As more information is accumulated, it becomes more useful. For example, recording the dates and weather conditions of Spring Peeper breeding congregations from year to year allows the observer to predict these activities in future seasons and note changes in populations.

There are various methods of keeping field records. A small, durable field notebook is simple and effective. Tablet computers and smartphones are good alternatives, especially with built-in GPS. Numerous programs exist for maintaining records and photographs. The North American Field Herping Association has an application for cell phones so that locality information and photos can be uploaded to a centralized database from the field (see Resources, this volume). The DNR will accept distribution records, including photos, through their Web site: http://www.dnr.state.mn.us/eco/mcbs/amphibian&reptile_maps.html. Locality

Red-bellied Snakes found under cover boards. Photograph by Minnesota Department of Natural Resources—Christi Spak.

Drift fence arrays. Photograph by Minnesota Department of Natural Resources—Carol D. Hall.

Turtle traps. Photograph by Minnesota Department of Natural Resources—Carol D. Hall.

information is best represented in latitude and longitude. These data can be obtained from a handheld GPS unit or cell phone.

Photos need to be well focused and close enough to see identifiable characteristics of the species. Photos do not have to be in natural habitat. Photos can be taken by any type of camera, including cell phones. If possible, a photo of the belly is very helpful.

Advanced field study may require marking individual animals for recapture studies. Mark and recapture programs are used to determine such things as territories, migration patterns, longevity, and population dynamics. These programs are best conducted by trained biologists with specific goals in mind. Indiscriminate marking of an animal does little to further research and may result in death or injury of the individual if the mark draws inordinate attention to it. Any study requiring the marking of animals should be coordinated with the DNR before the start of the program.

Various methods of marking employed by field biologists include painted-on numbers, shell notching (turtles), metal or

Good photograph for identification. Photograph by Minnesota Department of Natural Resources—Carol D. Hall.

Poor photograph for identification. Photograph by Minnesota Department of Natural Resources—Carol D. Hall.

plastic tags, and passive integrated transponders (PIT tags) (Ferner 2007). Radiotelemetry allows the researcher to relocate an animal and track its movements. A small radio transmitter powered by batteries is attached to or surgically implanted in the study specimen, and a receiver with an antenna is used to relocate the animal after its release. Radiotelemetry has provided natural history information about animals that would otherwise be unattainable.

Care of Captives

The fascination with amphibians and reptiles may compel a person to bring one home as a captive. For the sake of the animal, considerable forethought should go into this decision. When you remove an amphibian or reptile from the wild, you are taking on the responsibility for the welfare of that animal. Make sure the animal can be legally taken (i.e., it is a nonprotected species from nonrestricted public or private lands). Suitable escape-proof housing should be prepared before the collecting trip. Keep the numbers of captives to a minimum. Many individuals are nervous and do not settle down in captivity. Others may languish and refuse to eat or drink. Animals that do not adjust to captivity should be released where they were collected before their physical condition deteriorates to the point that their continued survival is unlikely. Long-term captives and animals of unknown origin should never be released in the wild. Sick or lethargic animals should not be released because they could introduce disease to the wild populations. These factors make buying captive bred animals for pets the best choice.

Providing a suitable cage environment and appropriate food are key requirements for successful keeping. Suitable cage warmth and lighting are necessary for animals to remain active and feed properly. Most species require a supplementary heat source during the cooler months. Provide an area of seclusion (such as a hide box) within the enclosure to help the animal better adapt to confinement. Cages and enclosures should be kept clean to reduce the buildup of disease organisms. Many species require live food such as insects, rodents, or other small animals. Providing a constant supply of these food items may be difficult or impossible. It is best not to feed live rodents to a caged snake, as the snake may be attacked by the mammal.

Pet stores that specialize in amphibians and reptiles sell live and frozen food, caging, and accessories. Informed store keepers can be helpful with husbandry problems, but it is best to rely on the advice of individuals thoroughly familiar with captive amphibian and reptile care. Numerous publications and Web sites provide reliable information about captive care of amphibians and reptiles. Additionally, the Minnesota Herpetological Society (see Resources, this volume) is a regional organization open to the public with many members knowledgeable in proper techniques of amphibian and reptile care.

Keeping venomous reptiles captive is highly discouraged. A person who maintains dangerous animals is legally liable if problems arise. Many municipal ordinances prohibit the keeping of such animals, and few neighbors or landlords accept or tolerate such practices. Minnesota's two venomous species are protected by law and cannot be kept by private parties.

Caring for and observing captive amphibians and reptiles can be rewarding and fascinating. Many youngsters have developed a lifelong interest in herpetology and a greater appreciation for wildlife from their first frog or snake, but the *best* place to observe and study amphibians and reptiles is in their natural environment.

CONSERVATION

Preventing extinction of a species is a compelling reason for the conservation of our native amphibians and reptiles. A portion of an ecosystem and its function is lost forever with extinction.

Unfortunately, humans have contributed to the rapid loss of many kinds of plants and animals throughout the world. Disrupting the balance of nature and destroying biodiversity are ongoing processes occurring at an accelerated pace throughout much of Minnesota, especially in the more populated areas. While no species have been officially extirpated from the state, the Massasauga is probably gone.

Habitat Loss

Destruction and loss of habitat are the greatest threat to amphibian and reptile populations. Without a home these animals perish. Amphibians and reptiles have restrictive home ranges, and they lack the ability to wander great distances to locate new living space. Also, their ability to adapt to habitat change is limited. Obviously a shopping center and parking lot destroy the habitat, but alterations such as mining, agriculture, and housing developments also have significant adverse impacts. Recreational activities of humans may indirectly disrupt the habitat and the normal activities of amphibians and reptiles. An example is the damage done to turtle nesting sites by unintentional trampling by boaters and campers along riverbanks. Mammalian predators (e.g., striped skunk, raccoon, and Virginia opossum) dig up turtle nests and consume the eggs. Since these mammals are attracted to human recreational areas in search of food, many more nests are found and destroyed in these areas.

Predated turtle nests, showing destroyed eggs, along Mississippi River. Photograph by Minnesota Department of Natural Resources—Carol D. Hall.

One of the main causes for the decline of many amphibians and reptiles is habitat fragmentation caused by human activities such as roads, agriculture, and urbanization. The effects of fragmentation include increased road mortality by vehicles and disruption of migrations to breeding sites and hibernacula caused by barriers to movement (Andrews, Gibbons, and Jochimsen 2008). Fragmentation also makes it easier for predators to find reptiles and amphibians.

Roads are a serious problem because they fragment habitat and invite basking on

Sand and gravel mine. Photograph by Minnesota Department of Natural Resources—Fred Harris.

warm days. Increased road kill can cause the local extirpation of a population. A Northern Leopard Frog population in western Hennepin County was highly reduced because the frogs could not get across a two-lane county road that separated the wintering site (Lake Independence) from the breeding habitat and summer sites (Baker Park Reserve) (Linck 2000). The mortality was high because the frogs were moving in early spring when it was still cool, their movements were slow, and the frog's defense is to crouch down and stay still. This tactic does not work well with motor vehicles.

Peat mine. Photograph by Minnesota Department of Natural Resources—Carol D. Hall.

Snakes and turtles also use sitting still as a defensive tool. Studies have shown that some people will try to avoid turtles on the road, but a larger percentage will swerve to hit a snake (Ashley, Kosloski, and Petrie 2007).

The consequence of habitat loss is especially critical for rare species. If valuable natural areas can be identified early, it is possible to set aside and protect prime habitat for the uncommon species. The DNR has a variety of habitat protection and inventory programs, including the Scientific and Natural Areas Program, the Nongame Wildlife Program, and the Minnesota Biological Survey. Together these programs have placed increased emphasis on inventorying the state's biological resources and protecting critical habitats. Wetland restoration efforts by both state and federal resource agencies also have provided benefits to Minnesota's amphibians and reptiles. Nongovernmental conservation organizations, especially The Nature Conservancy, have further contributed toward statewide habitat protection goals. With strong citizen support, these and other conservation efforts will help ensure the preservation of Minnesota's amphibians and reptiles.

Road-killed Blanding's Turtle. Photograph by Minnesota Department of Natural Resources— Eric Halbur.

Road construction and grading. Photograph by Minnesota Department of Natural Resources.

Snapping Turtle enters onto the road. Photograph by Allen Blake Sheldon.

Turtle-crossing sign. Photograph by Barney Oldfield.

Pollution

Unfortunately, little information is available about the susceptibility of amphibians and reptiles to environmental contaminants. An entire population or species could perish before the cause of the tragedy is discovered. Pesticides accumulate in prey species and over time may poison predator species as well. The loss of Spring Peepers in the Twin Cities metropolitan area may be related to the intensive use of chemicals in the urban environment as well as to significant habitat loss. Research has shown that the water in ponds contaminated with acid rain can prevent successful breeding in some amphibian species. Environmental toxicological research is urgently needed to develop effective methods of cleaning up contaminated areas. We must find ways to greatly reduce obvious and insidious forms of chemical pollution before it is too late for a number of our amphibians and reptiles.

Malformed Eastern Tiger Salamander. Photograph by Allen Blake Sheldon.

Malformed Northern Leopard Frog. Photograph by Allen Blake Sheldon.

AMPHIBIAN DECLINES AND DISEASES

In summer 1995 a group of students from New Country School found a large number of Northern Leopard Frogs with missing, deformed, and extra limbs (Souder 2000). This discovery was preceded by the observation of another deformed frog population in 1993 in Granite Falls along the Minnesota River (Helgen 1997), which led to an initial investigation by the Minnesota Pollution Control Agency (MPCA) (Helgen et al. 1998). The MPCA work expanded to statewide (Helgen et al. 2000; Rosenberry 2001) and national (Souder 2000) surveys and research into amphibian deformities and their cause.

Research on deformities in Minnesota looked at numerous possible causes, including UV radiation (Blaustein and Belden 2005), chemical contamination (Hoppe 2000), predation (Sessions 2003), and parasites (Johnson and Lunde 2005). Numerous conferences and symposia on the potential causes were held (Souder 2000; Lannoo 2008), several books on the topic were published, and a national clearinghouse on amphibian deformities, the North American Reporting Center for Amphibian Malformations (NARCAM), was formed (Johnson, Fowle, and Jundt 2000).

After seven years of research and surveys, the researchers still did not have a single cause, but chemical pollution and parasites were determined to be the main factors (Souder 2002, 2005). During the past five to six years, reporting of deformed frogs has greatly diminished in Minnesota and across the country. Most research has ended, and the NARCAM was closed after 2006. Frog deformities are still occurring, but the public and most researchers have become uninterested. Helgen (2012) reported on continued issues and ongoing research in Minnesota and across the country.

One area of research still being conducted is the effect of atrazine on reproduction in frogs (Hayes et al. 2002, 2011). Hoppe (2000) found several ponds with Northern Leopard Frogs that did not initiate breeding. The atrazine disrupts hormone levels in the frogs, causing the males to feminize and not breed. The males will actually start developing eggs in their testes.

As the deformed frog issue was quieting down, concern about *Batrachochytrium dendrobates* (chytrid) fungus and *Ranavirus* was growing. The fungus and virus had been tied to the massive die-offs of frogs in Central America and parts of the western United States (Greer, Berrill, and Wilson 2005). The U.S. Geological Survey's Amphibian Research Monitoring Initiative (ARMI) has been doing surveys across the state (Sadinski and Roth 2009). Rodriquez et al. (2009) published studies on chytrid in Lake Itasca State Park, and Uyehara, Gamble, and Cotner (2010) did a survey of *Ranavirus* at the same location The fungus and virus can be spread from site to site on researchers' and visitors' boots and shoes as they move from wetland to wetland (Schloegel et al. 2010). Chytrid fungus has been associated with frog die-offs at two sites in Minnesota (L. Brown, pers. comm.). Researchers now follow special procedures that include drying and disinfecting boots and field equipment between sites (Kingsbury and Gibson 2012). Nature watchers and field herpers should also be careful not to spread the fungus by following the same procedures as the researchers. Equipment can be cleaned with a bleach solution, or equipment can be allowed to dry for several days between visits.

Mudpuppy die-offs in lakes of west-central Minnesota have puzzled researchers. Chytrid fungus and *Ranavirus* have not been found in the specimens (K. Larson, pers. comm.).

HARVESTING PRESSURES

The commercial harvest of Minnesota's amphibians and reptiles varies by group and species. Current state regulations are in place to manage the harvest of turtles and frogs, but there are no regulations for snakes, lizards, and salamanders, except for the Minnesota species listed as endangered or threatened (Minn. Rules, chapter 6134).

Laws pertaining to turtles were revised several times between 2001 and 2004. The revisions designated Snapping, Painted, and Spiny Softshell Turtles as commercial species, and only these three species can be commercially harvested or sold. The new laws include a recreational trapping license that allows

Commercial turtle trap.
Photograph by Tony Gamble.

Softshell turtle tracks in mud.
Photograph by Tom Jessen.

an individual to set up to three traps. Harvested turtles cannot be sold and may be used only for personal consumption. The laws also limit commercial turtle trapping licenses to existing license holders. Since 2001, the number of trappers has dropped from 118 to 27 in 2011. This number will continue to decrease as trappers retire. The actual harvest of turtles has also decreased.

Painted Turtles are mainly harvested for biological supply businesses and the pet trade. Minnesota harvest peaked at over 60,000 turtles per year prior to 2000 but has dropped to an average of 4,000 turtles per year in the past five years (Larson 2012). Snapping Turtles are harvested for their meat. The number harvested peaked at 5,000 turtles per year and has dropped to approximately 1,000 turtles per year (Larson 2012). Spiny Softshells are harvested for their meat, but the number harvested has always been low, averaging fewer than 200 turtles per year (Larson 2012). These three species are also legally harvested recreationally by anglers with a fishing license. There are set seasons and size limits. The recreational harvest is not tracked by the DNR.

Frogs are commercially harvested for fishing bait by commercial bait suppliers. The bait harvest is mainly limited to

Commercial frog harvesters, 1915. Photograph courtesy of the Minnesota Historical Society.

Northern Leopard Frogs. Harvest of frogs for purposes other than bait requires a special permit. The DNR does not track the number of frogs harvested for bait, and the harvest for purposes other than bait is minimal (N. Vanderbosch, pers. comm.).

Salamanders, especially Eastern Tiger Salamanders and Mudpuppies, are commercially harvested for bait and biological supply houses. There are no regulations on their harvest. The DNR is currently doing statewide surveys to assess the status of the Mudpuppy population.

Gartersnakes are harvested in the northwest portion of the state for the biological supply and pet trade (J. Konrad, pers. comm.). The DNR does not have any regulations for nonthreatened snakes. Research is needed to determine if harvesting is negatively affecting Gartersnake populations.

The pet trade is a minor part of the commercial harvest, outside of the species previously mentioned. The main targets of the pet trade are large snakes and turtles. Several species, including Wood and Blanding's Turtles, Timber Rattlesnakes, and Western Ratsnakes, that had previously been sought after by collectors are now protected by the state endangered species law (Minnesota DNR 2013). The species currently unprotected include Gophersnakes, Western Foxsnakes, Plains Hog-nosed

Snakes, and Milksnakes. The collection of these species should be monitored to detect any negative effects on the populations.

PERSECUTION

Because of fear, hate, and ignorance, numerous snakes are deliberately killed each year. Snake haters dispatch snakes on sight, assuming they are ridding the world of dangerous pests. Uninformed people often kill harmless species because they mistake them for venomous species. Killing snakes is seldom justified; most snakes are beneficial because they eat numerous rodents and insects. Even the life of venomous snakes can be spared if a person knowledgeable about the species is contacted. Call the local animal control office or conservation officer for assistance in contacting knowledgeable authorities.

Throughout history, humans have singled out and intentionally persecuted certain species. For more than fifty years, Minnesota had an active rattlesnake bounty. It was administered by local township boards in several southeastern counties. For the years 1967 through 1982, Houston County paid bounties for 28,685 rattlesnakes. Only 191 bounties were paid in Minnesota in 1987. A Minnesota statute removing the authority to pay bounties for rattlesnakes was signed into law in 1989. Bounty systems have repeatedly been shown to be biologically unsound as well as difficult to monitor and administer.

COMMON AMPHIBIAN AND REPTILE PROBLEMS

A number of common amphibian- or reptile-related "problems" are of concern to the general public. Some of the problems deal with the well-being of the animal, but usually the well-being of the human is at issue. Many of the problems are easily solved by exclusion of the amphibian or reptile from areas where people do not want them. The problems or situations described in this chapter address the most common human interactions with amphibians and reptiles. If other situations arise, consult the DNR's Web site, or contact a regional nongame wildlife biologist at the DNR. They can provide additional information or direct you to someone who can help. This section does not address problems that might arise from the captive maintenance of reptiles or amphibians. Those questions can be answered by members of the Minnesota Herpetological Society.

Snakes in the House or Garage

A frequent problem encountered in the spring is the unwanted appearance of snakes, especially gartersnakes, basking on the patio or front steps. Homeowners also frequently complain about snakes in their basements or garages. Both of these problems are usually due to cracks or holes in the house or foundation of the house that allow snakes access. Gartersnakes hibernate behind concrete steps or patios because these spaces provide a warm (above freezing) place to spend the winter. Snakes in a basement or garage may also be looking for a place to overwinter. Snakes will also enter a house or garage to search for food, specifically mice.

This problem can be eliminated by sealing the cracks between the house and stoop or patio and filling holes or cracks in the foundation. Most snakes can fit through a crack ½ inch wide (1 centimeter). Another way to discourage snakes is to eliminate their food source. Mice like to nest in wood piles stacked adjacent to or in the house or garage. Mice also come through the same cracks as the snakes. Mice can be controlled by using traps or poison baits. Standard procedures and chemicals used by pest control companies are generally ineffective for dealing with snakes.

Snakes in the Yard

Snakes are regularly found in yards in rural areas because they are attracted to cover and food resources found near yards. The easiest way to discourage snakes is to make your yard unattractive to them. These measures include removing wood piles from the vicinity of buildings and use areas, trimming shrubs and trees to create a space of at least 6 inches between the ground and first branches, and keeping the grass short throughout the yard. The only way to absolutely keep snakes out is by fencing. Snake-proof fencing can be made by modifying a normal chain-link, picket, or split-rail fence by attaching an 18- to 24-inch-high piece of hardware cloth to the outside of the fence.

Bury the bottom of the hardware cloth 2 to 4 inches into the soil. Snakes will tend to travel along a fence rather than go over a fence (Moriarty 1999).

Salamanders in the Basement

Tiger salamanders are commonly found in older basements made of stone or cement block where there are spaces between the stones or blocks. They are also commonly found in wells and cisterns in rural areas. Generally, the salamanders are searching for a place to hibernate. If they do not find a moist site, they will desiccate and die. Salamanders are totally harmless to people and will not cause any damage, and they can be controlled by sealing the cracks in the foundation. The solution for keeping both snakes and salamanders out is the same. Sealing cracks is also a good way to save energy in your house and cut down on drafts. Salamanders and frogs can also become trapped in window wells on the outside of the foundation. Window well covers will keep these and other wildlife from becoming trapped.

Turtles Nesting in the Yard

More and more houses are being built along lakes and wetlands. These shorelines are the same place where turtles have been laying their eggs for centuries. The turtles will continue to nest in the same area even if it has been changed into a manicured lawn or child's sandbox. Species commonly encountered in yards are Painted Turtles, Snapping Turtles, and Blanding's Turtles. The only turtle that may cause a safety problem is the Snapping Turtle because of its size and temperament.

The best way to deal with these turtles is to leave them alone. If a turtle is seen nesting in a yard, it should be observed, and the nest could be marked when egg laying is completed. Marking will keep the nest from being accidentally disturbed by humans, and hatchlings can be observed when they later emerge. The nest can be protected with a wire cage. Examples of wire cages are available from the DNR. If you do not want a nest in your yard, you must return the turtle to its pond before it lays its eggs. You may have to do this several times before she will go to a different area to nest. A fence around the perimeter of the yard will also keep turtles out. The fence has to go all the way to the ground and be sturdy or the turtle will crawl under the fence.

If a turtle does lay her eggs before she can be moved, do not dig up the nest. Turtle eggs are easily killed by inexperienced handling. We do not recommend the transplanting or artificial incubation of turtle eggs.

Snapping Turtles Eating Ducklings

The Snapping Turtle is a predator and scavenger by nature. It mainly eats fish, carrion, and vegetation. They will occasionally take ducklings. This is a normal occurrence in nature. In rare situations a turtle population becomes overabundant in a lake or pond and causes increased duckling mortality. Turtles can be controlled by trapping. Commercial turtle trappers can be hired to remove the turtles, but they can only keep turtles over 12 inches long, so continual control may be necessary.

Salamanders, Frogs, and Toads in the Road or Yard

Spring rains initiate the movements of amphibians in search of breeding ponds. In some areas these migrations consist of thousands of individuals. When these migrations intersect a road, the outcome can be a massacre. The same situation occurs in the fall when young frogs and salamanders are moving to hibernating sites. At some localities the migrations can be so large that roads become slick with dead animals that are hit by vehicles.

Protection of amphibian migrations has recently become a conservation issue. In some areas in Europe and the United States tunnels have been constructed to direct the animals to pass under the road. In Massachusetts, some roads are closed during salamander migrations. Minnesota has installed signage in some areas and has begun to redesign stream crossings for safer passage of wildlife, including reptiles and amphibians.

Homeowners are sometimes upset by having numerous frogs and salamanders in their lawns and driveways when the movement corridor of an amphibian migration goes through a yard or residential area. However, the animals will not hurt the property, and they will be gone within several days. Grass mowing and other yard activities should be curtailed during the migration.

AMPHIBIAN AND REPTILE IDENTIFICATION

How to Use This Book

A principal objective of this book is to enable the user to identify Minnesota's amphibians and reptiles. Making an identification will likely require having the animal in hand or located where it can readily be observed. After an identification is made, the reader can become more familiar with that species' distribution, habitat requirements, and life history.

To make an identification, refer to the circular keys at the end of this chapter and decide which taxonomic group the animal belongs to (i.e., salamanders, frogs, turtles, lizards, or snakes). Then, go to the appropriate key to determine the species. Examine the photographs, and carefully read the descriptions of the species that resemble the one in question to support the identification. Checking the range map can be helpful, but do not rule out an identification by distribution alone because it may be a new county record.

If an animal cannot be identified, expert help is available through the Minnesota Herpetological Society, the Bell Museum of Natural History, the DNR's Nongame Wildlife Program, or a local college's biology department. An unusual animal may be an escaped captive or may have been transported to Minnesota by way of truck or train. Several commonly reported species are the Eastern Box Turtle *(Terrapene carolina)*, Ornate Box Turtle *(Terrapene ornata)*, and Green Treefrog *(Hyla cinerea)*. Several species are found in adjacent states near the Minnesota border, and it is possible that they may someday be verified as Minnesota residents. Brief descriptions and photographs of these neighbors and potential inhabitants are found in the section "Species of Possible Occurrence."

Description of Species Accounts

Each species account is divided into subsections, which are described below. Color photographs and a Minnesota range map are included with each account. The description and photographs aid in species identification, while the remainder of the material provides useful and interesting information concerning the natural history of the animal.

Name: Each species has a common and a scientific name. Generally people are comfortable using the common name of an animal; however, confusion arises when one person refers to a particular kind of turtle as a "mud turtle" and someone else calls it a "painted turtle." To avoid confusion, scientists assign a two- or three-part Latin or Greek name to each species when it is first described. This name is based on its relationship with similar species and is recognized as that particular species worldwide.

Common and scientific names used in this book are taken from *Scientific and Standard English Names of Amphibians and Reptiles of North America North of Mexico* (Crother 2012). This is the official name list for all the national herpetological organizations. There have been a number of changes since the publication of *Amphibians and Reptiles Native to Minnesota* (Oldfield and Moriarty 1994). These changes are noted in the "Remarks" subsection of the species accounts.

Photographs: Photographs were selected to provide a typical and overall view of each species and to demonstrate key diagnostic features. Additional photographs are used to demonstrate key features, unusual color variations, juvenile patterns, behavior, or habitat.

Map: The base map is a map of Minnesota counties modified so that large counties are subdivided (W. S. Smith 2008). The subdivisions provide a better representation of distribution for the northern portion of the state. The distribution maps are a composite of museum records, literature reports, and sightings. The museum records are mapped as pre-1960 (◒), post-1960 (●), and literature and sighting records (○). The maps are up-to-date as of January 2012. A Minnesota map with county names and a legend of the distribution symbols can be found following the table of contents.

The majority of the museum records are from the Bell Museum of Natural History. Other museums with records of Minnesota specimens include the Academy of Natural Sciences, American Museum of Natural History, Carnegie Museum, George Mason University, Milwaukee County Public Museum,

Museum of Comparative Zoology (Harvard), Nebraska State Museum, Royal Ontario Museum, Science Museum of Minnesota, U.S. National Museum (Smithsonian), University of Colorado, University of Kansas, University of Michigan Museum of Zoology, and the University of South Dakota. Vouchers at museums other than the Bell Museum of Natural History and the Science Museum of Minnesota were not examined unless they represented a range limit or rare species. The Minnesota Biological Survey has been the largest source of records over the past 20 years.

Literature reports were compiled by reviewing all of the Minnesota literature (Moriarty and Jones 1988; Moriarty 2006) and other pertinent publications. Sight records were collected through the Minnesota Nongame Wildlife Program, Minnesota Herpetological Society, colleagues, and other interested persons. Records included in the distribution maps are from credible sources.

If you find a new county record or come across a species not known from a county since 1960, please contact the Bell Museum of Natural History or the Minnesota Nongame Wildlife Program (contact information is provided in Resources). To be an acceptable record, the actual specimen or a recognizable photograph accompanied with a detailed locality description must be provided.

Description: This section gives a description of a typical adult found in Minnesota followed by information about sexual differences, subadult color variations, and average and maximum sizes. Generally, measurements for salamanders, lizards, and snakes are given as total lengths (nose tip to tail tip). Anuran measurements are given as the length from snout to vent (tip of nose to vent), and turtle measurements are given as a straight line from the front edge of the carapace to the rear edge (not following the curve of the shell). Total lengths and snout–vent lengths are given for lizards since they frequently lose their tails, and the length given for rattlesnakes does not include the rattle because of its propensity for breaking. Finally, characteristics are noted that distinguish this species from others that might be confused with it.

Distribution: A brief description of the North American range of the species is given followed by its range in Minnesota. Maps of the entire North American distribution of Minnesota species can be found in Conant and Collins (1998).

Habitat: Many amphibians and reptiles have specialized living requirements; thus, each species' preferred habitat is described. Because hibernating requirements are directly related to habitat, information concerning hibernation sites is also included in this section.

Life History: This section provides information on periods of seasonal activity, daily activity, home range, food preferences, courtship and breeding behavior, egg development periods, number of young, defense strategies, predators, and general natural history.

Since frogs and toads have distinctive calls that are useful in identification, their vocalizations are described in this section. Written descriptions are helpful, but to become competent in recognizing calls, you may need to track down calling anurans or listen to recordings. Recordings that are useful for learning the breeding calls of resident frogs and toads can be found in Elliot, Gerhardt, and Davidson (2009) (see Resources).

The information in this book concerning natural history comes from the authors' experience, knowledgeable field personnel, and various references, which are listed in Literature Cited.

Remarks: This section provides additional information regarding a species that is pertinent or interesting, such as unusual behavior and longevity. When applicable, subspecies are described, and nonstandard common names known to the authors and name changes since Oldfield and Moriarty (1994) are provided. Also noted when applicable is a species' legal status as state endangered, threatened, or of special concern and its listing as a Species of Greatest Conservation Need.

Keys to Amphibians and Reptiles Native to Minnesota

The following keys are designed to assist in identifying the amphibians and reptiles found in Minnesota. The circular format allows you to easily move through the dichotomous (two-part) system. The first key will place you in one of the five major groupings: salamanders, frogs, turtles, lizards, and snakes. These keys are designed to be used with adult specimens (Keys 1–6) or aquatic amphibian larvae (Keys 7–8).

Start at the center of each key, and work to the outer ring. Drawings of certain features are referenced in the keys and found in Figures 1–6. When a potential identification has been made, refer to the photograph of the species in its account. If the photograph does not match the species, then back up to the previous ring and try again.

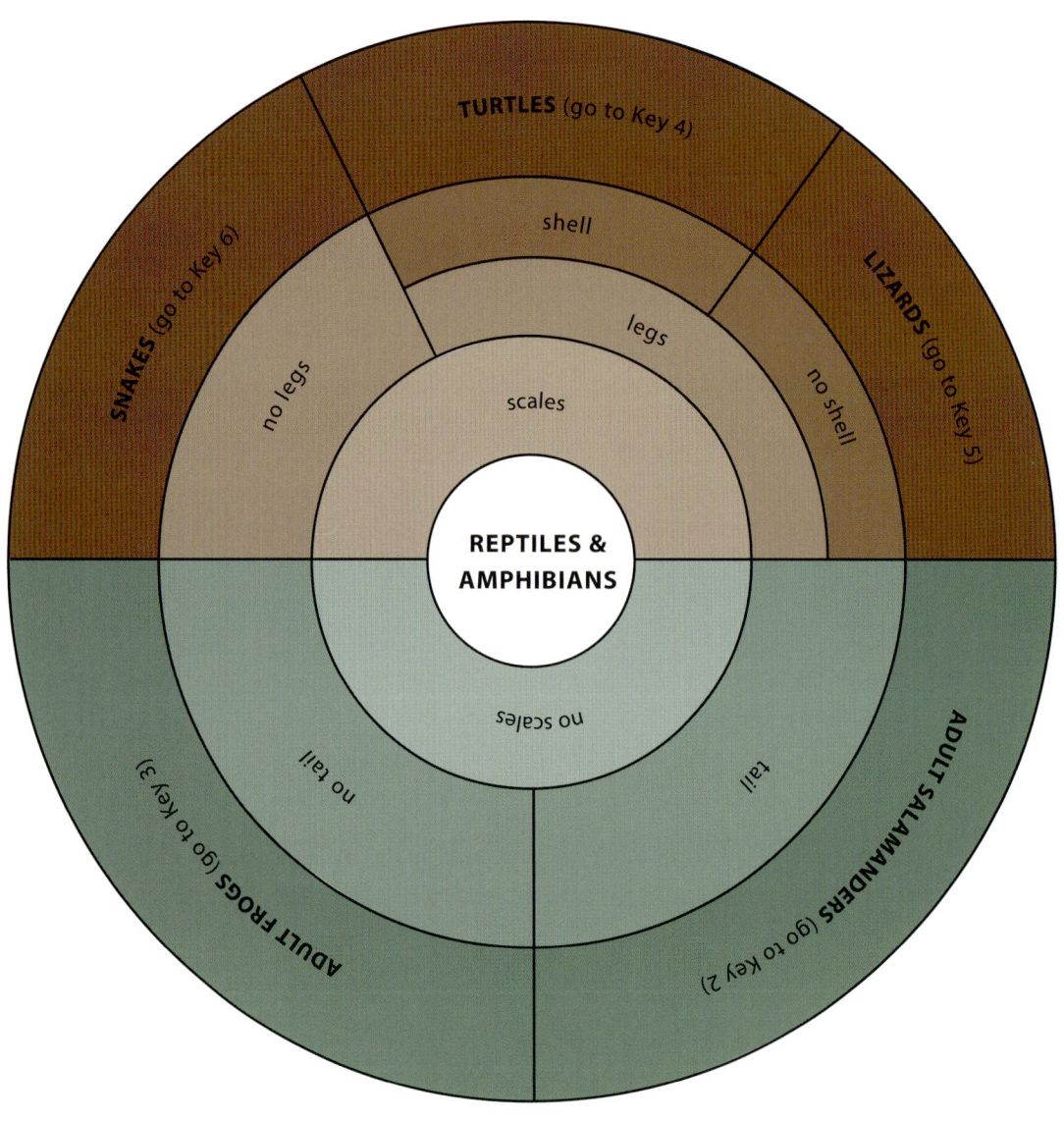

Key 1

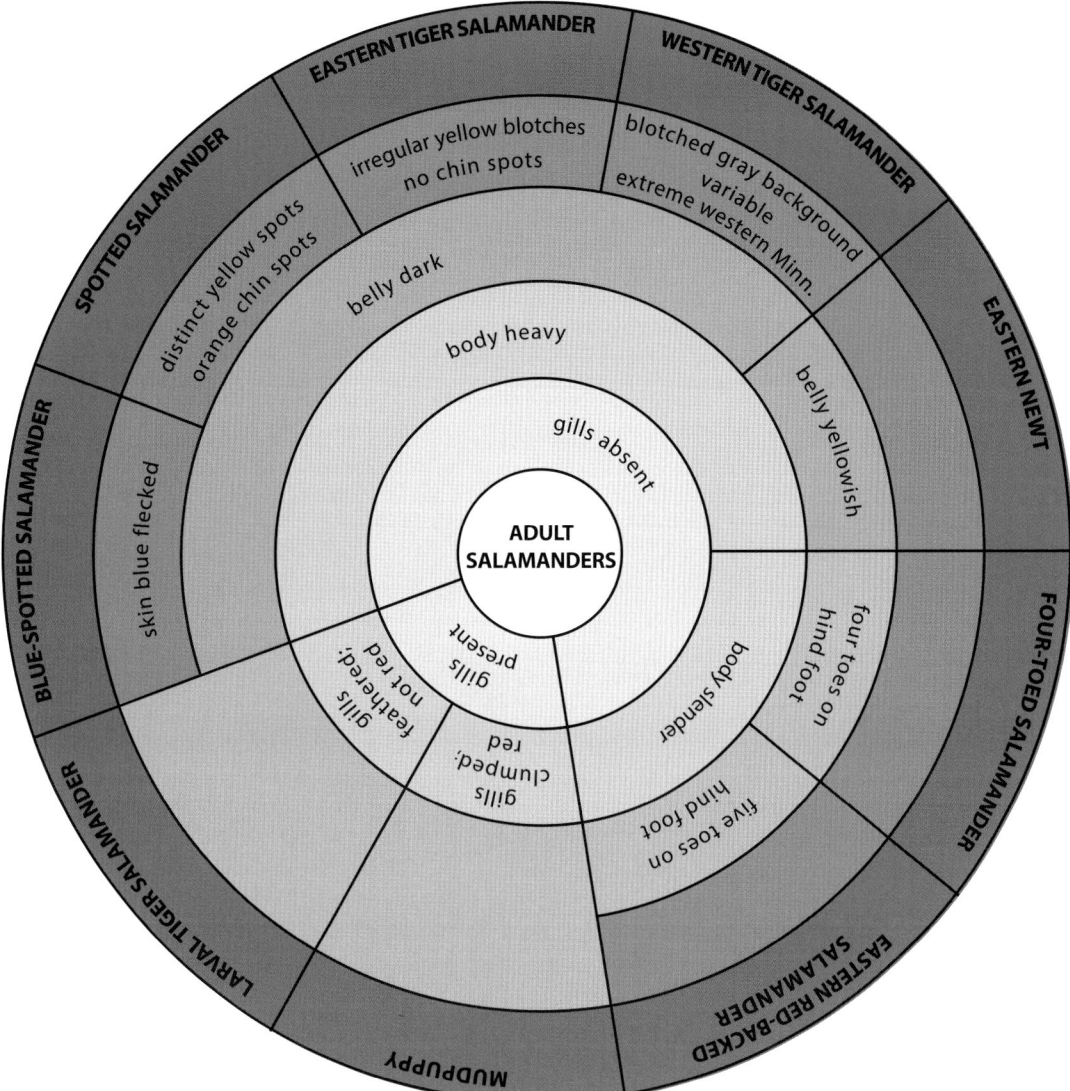

Key 2

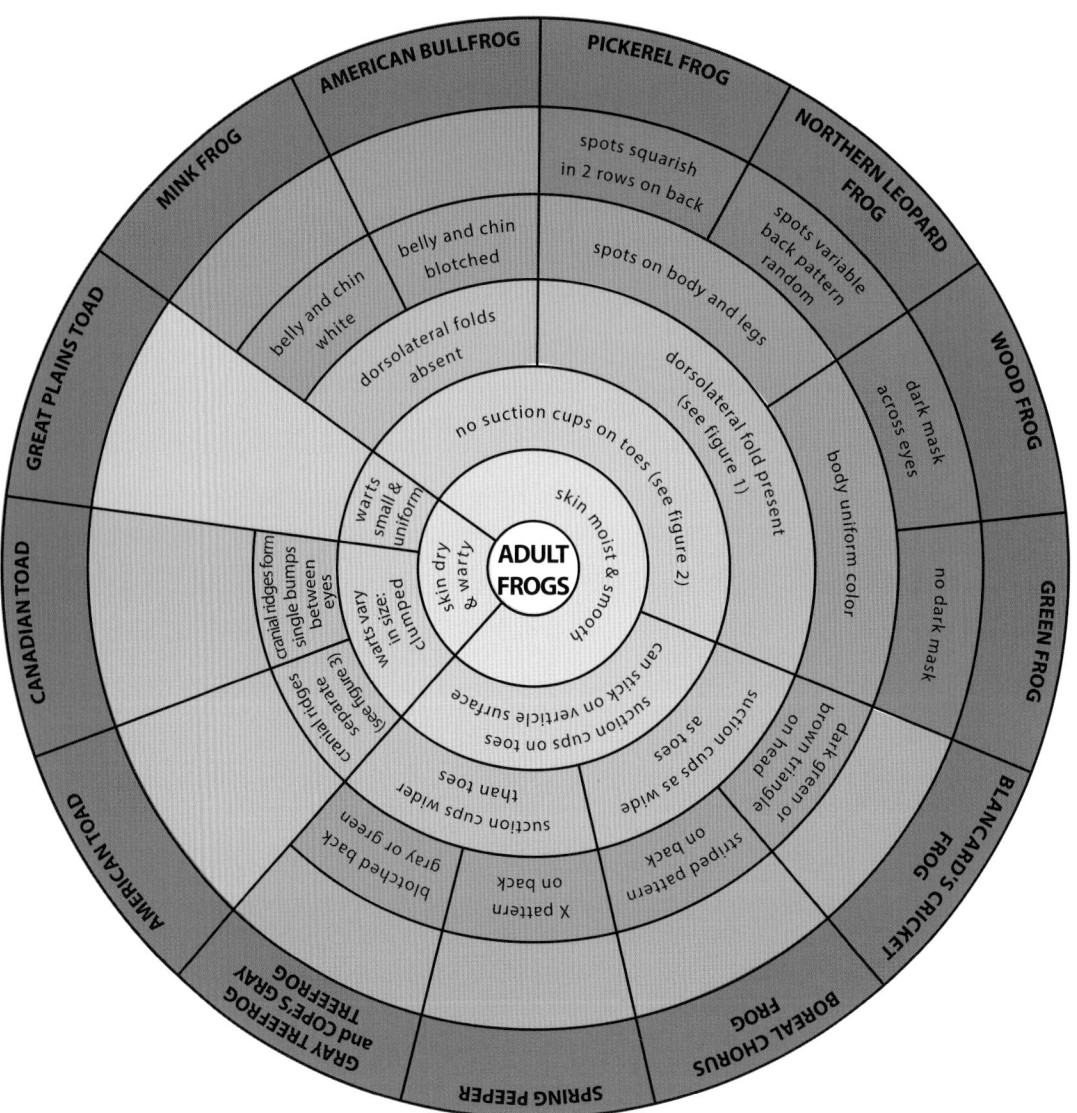

Key 3

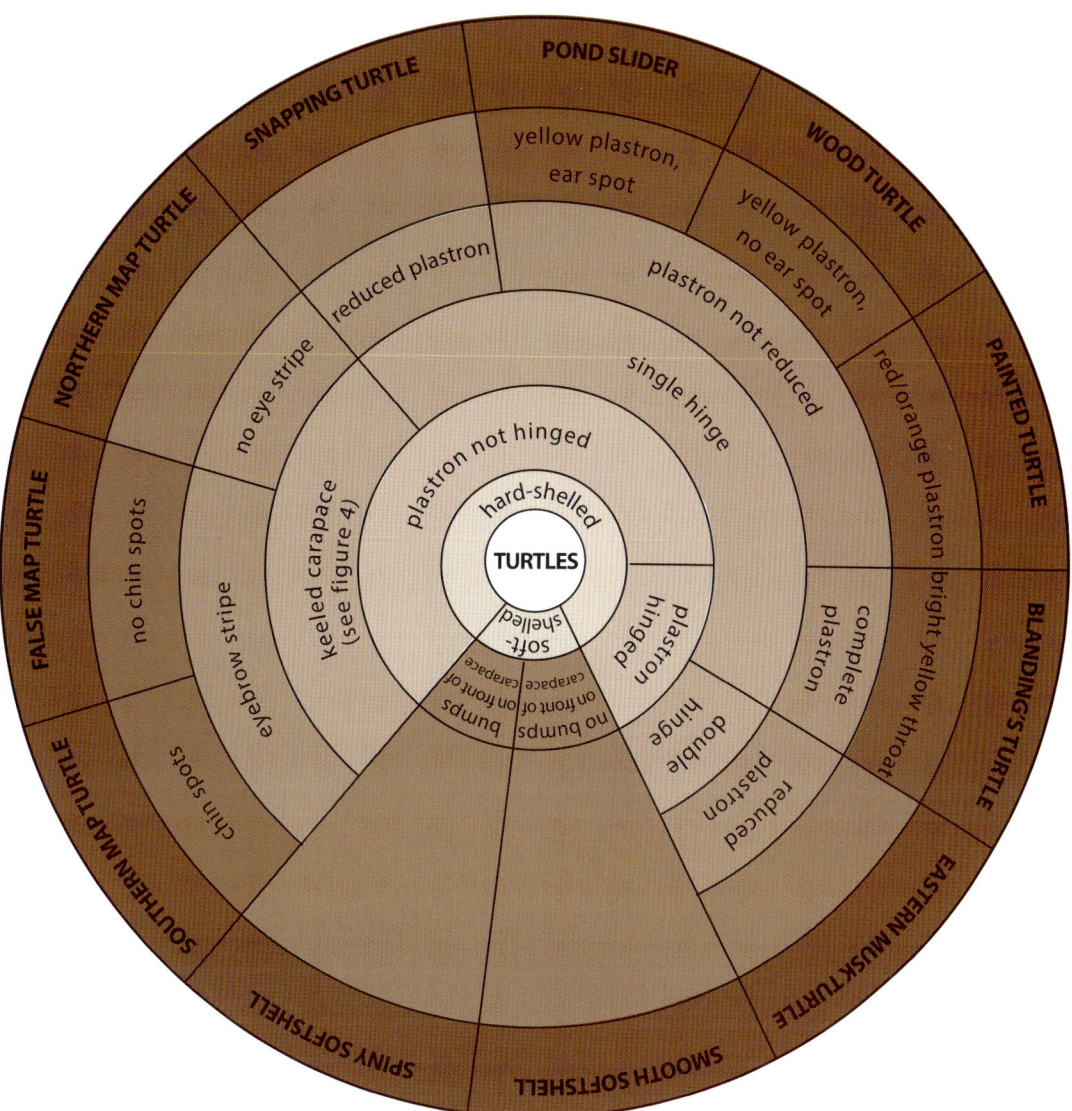

Key 4

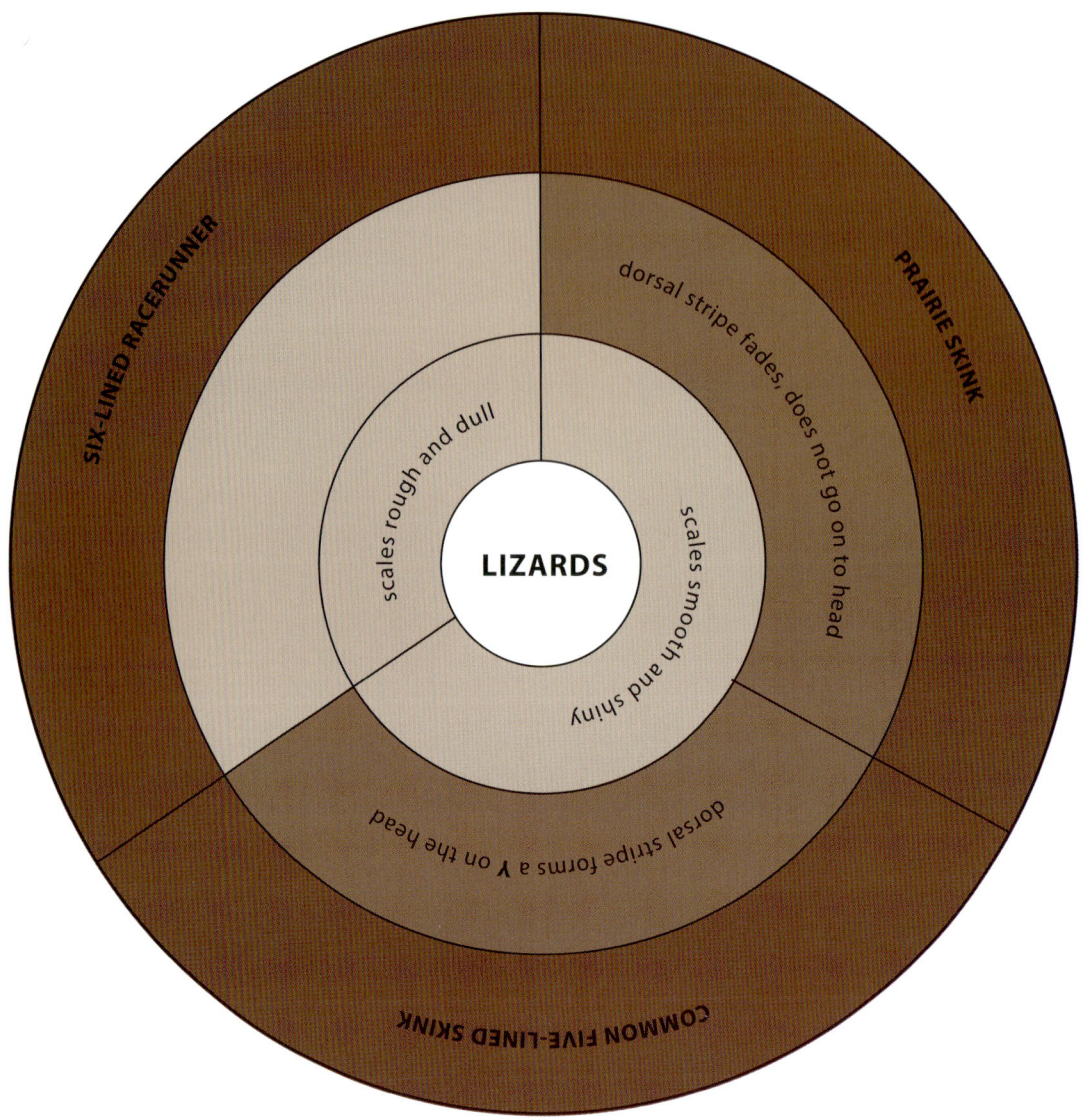

SIX-LINED RACERUNNER

PRAIRIE SKINK

dorsal stripe fades, does not go on to head

scales rough and dull

scales smooth and shiny

LIZARDS

dorsal stripe forms a Y on the head

COMMON FIVE-LINED SKINK

Key 5

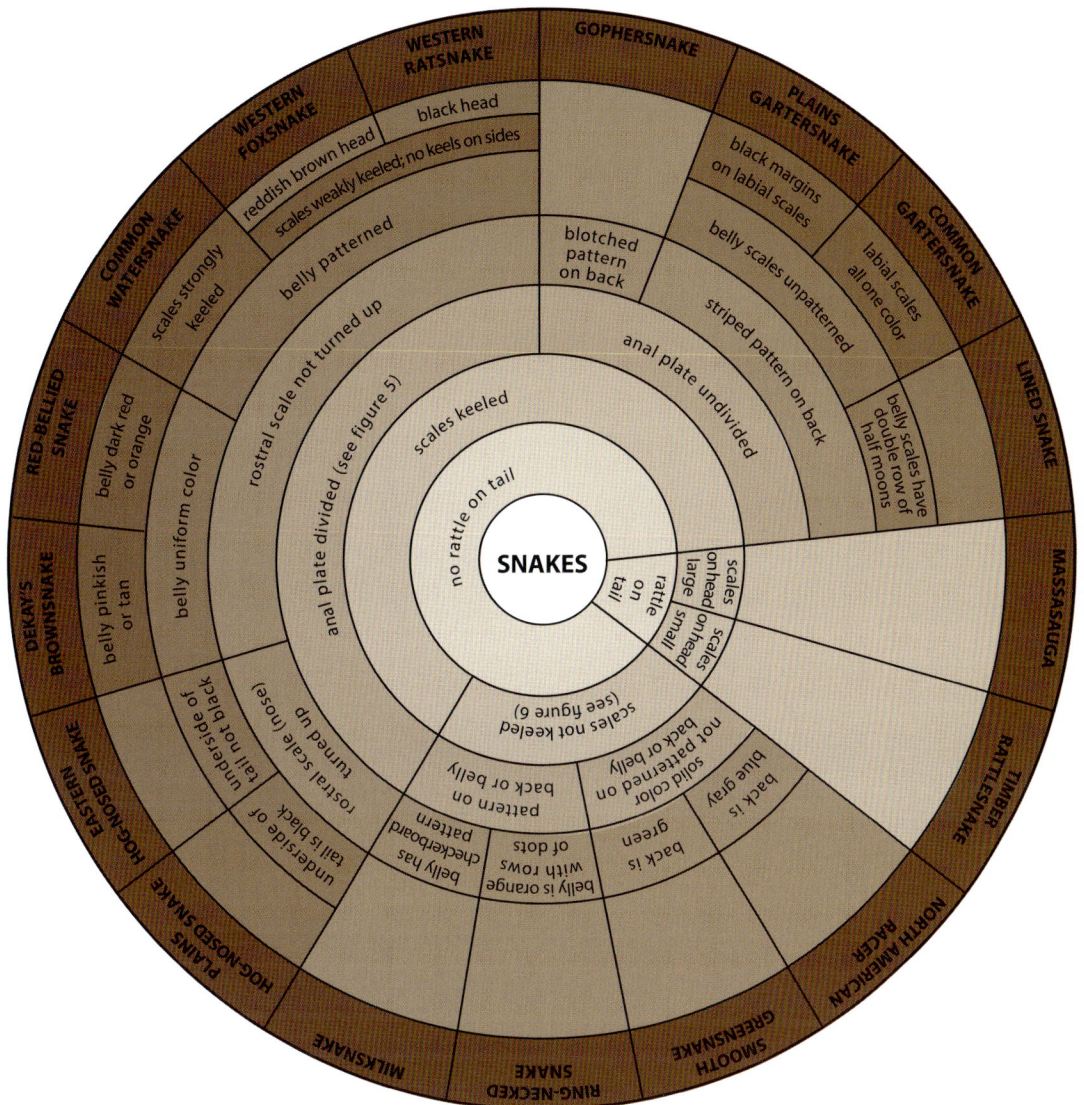

Key 6

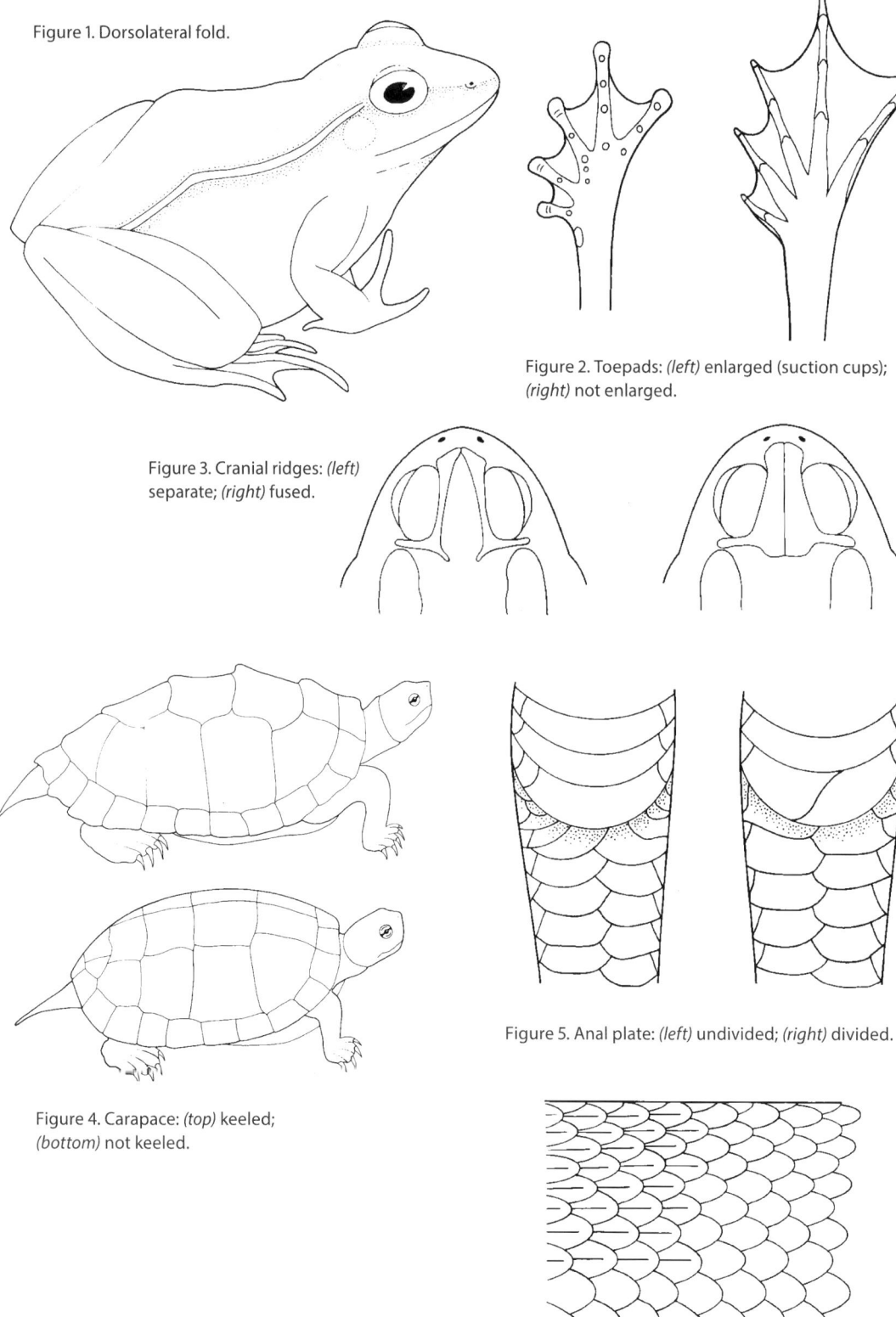

Figure 1. Dorsolateral fold.

Figure 2. Toepads: *(left)* enlarged (suction cups); *(right)* not enlarged.

Figure 3. Cranial ridges: *(left)* separate; *(right)* fused.

Figure 5. Anal plate: *(left)* undivided; *(right)* divided.

Figure 4. Carapace: *(top)* keeled; *(bottom)* not keeled.

Figure 6. Scales: *(left)* keeled; *(right)* not keeled.

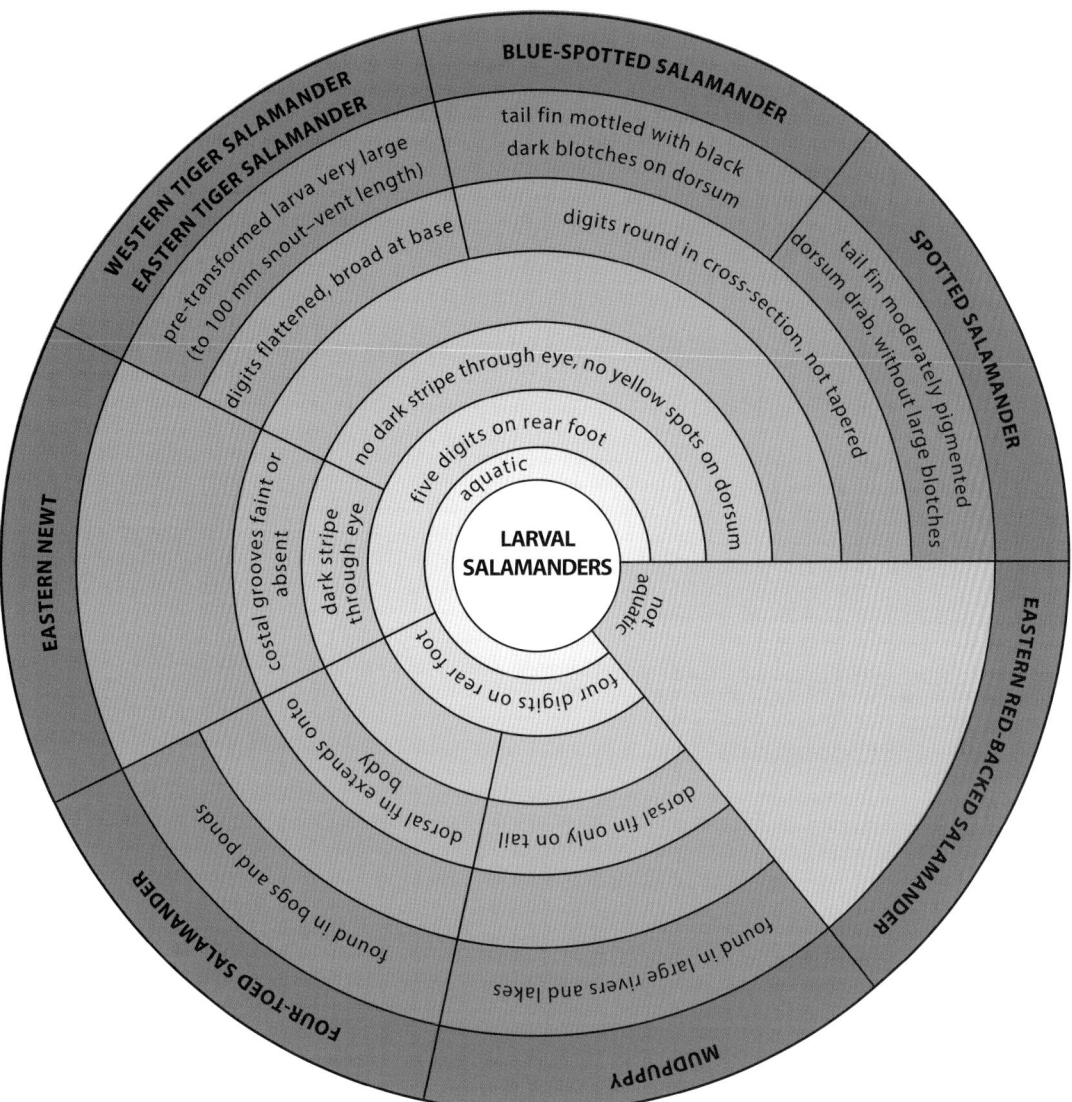

Key 7

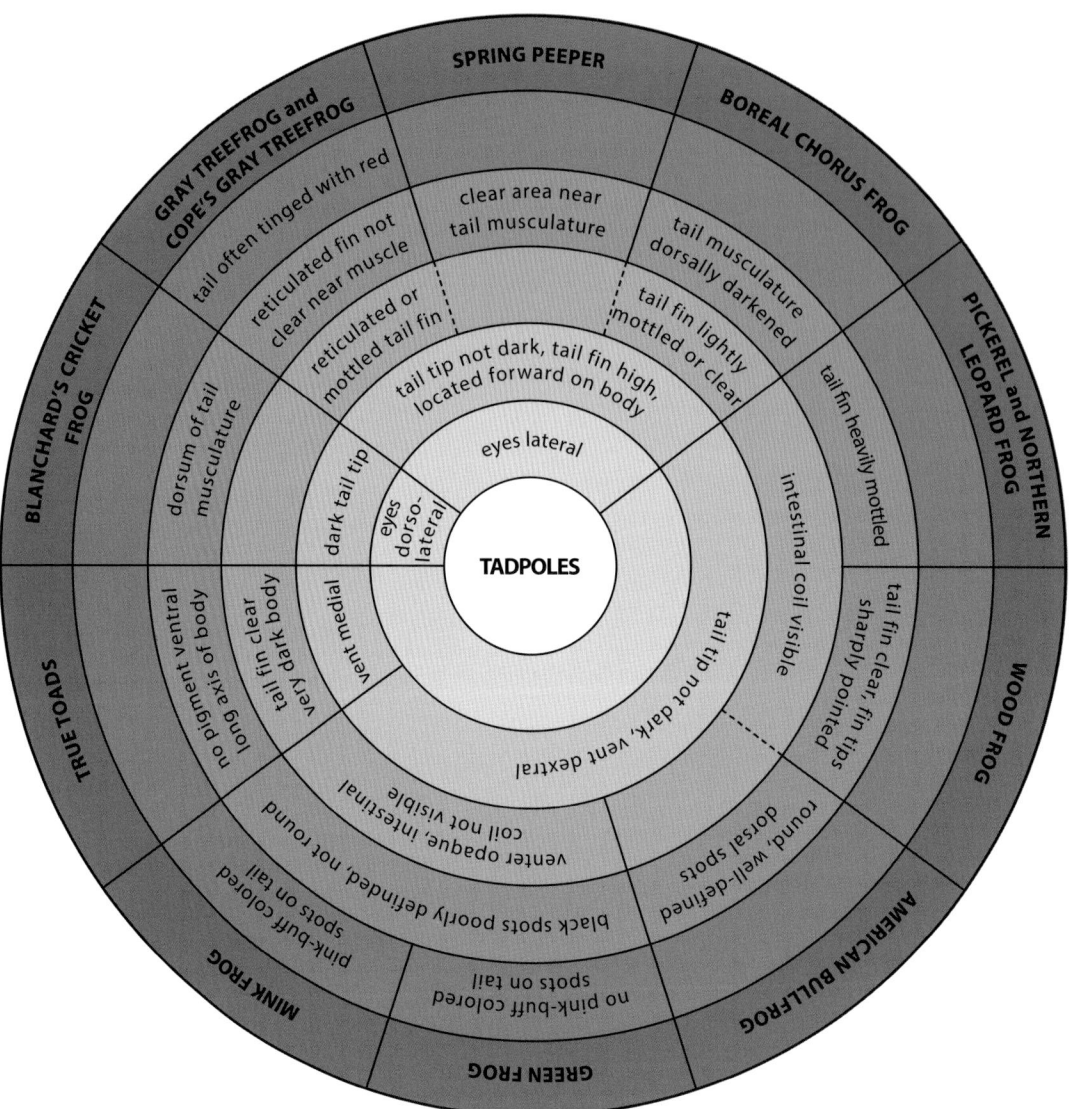

Key 8

Checklist
of the Amphibians and Reptiles in Minnesota

Names follow B. I. Crother, ed. (2012), *Scientific and Standard English Names of Amphibians and Reptiles of North America North of Mexico*, SSAR Herpetological Circular 39, 1–92.

FROGS

Acris Duméril and Bibron, 1841—Cricket Frogs
  A. blanchardi Harper, 1947—Blanchard's Cricket Frog

Anaxyrus Tschudi, 1845—North American Toads
  A. americanus (Holbrook, 1836)—American Toad
    A. a. americanus (Holbrook, 1836)—Eastern American Toad
  A. cognatus (Say, 1823)—Great Plains Toad
  A. hemiophrys (Cope, 1886)—Canadian Toad

Hyla Laurenti, 1768—Holarctic Treefrogs
  H. chrysoscelis Cope, 1880—Cope's Gray Treefrog
  H. versicolor LeConte, 1825—Gray Treefrog

Lithobates Fitzinger, 1843—American Water Frogs
  L. catesbeianus (Shaw, 1802)—American Bullfrog
  L. clamitans (Latreille, 1801)—Green Frog
  L. palustris (LeConte, 1825)—Pickerel Frog
  L. pipiens (Schreber, 1782)—Northern Leopard Frog
  L. septentrionalis (Baird, 1854)—Mink Frog
  L. sylvaticus (LeConte, 1825)—Wood Frog

Pseudacris Fitzinger, 1843—Chorus Frogs
  P. crucifer (Wied-Neuwied, 1838)—Spring Peeper
  P. maculata (Agassiz, 1850)—Boreal Chorus Frog

SALAMANDERS

Ambystoma Tschudi, 1838—Mole Salamanders
    A. laterale Hallowell, 1856—Blue-spotted Salamander
    A. maculatum (Shaw, 1802)—Spotted Salamander
    A. mavortium Baird, 1850—Western Tiger Salamander
        A. m. diaboli Dunn, 1940—Gray Tiger Salamander
        A. m. melanostictum (Baird, 1860)—Blotched Tiger Salamander
    A. tigrinum (Green, 1825)—Eastern Tiger Salamander

Hemidactylium Tschudi, 1838—Four-toed Salamanders
    H. scutatum (Temminck and Schlegel in Von Siebold, 1838)—Four-toed Salamander

Necturus Rafinesque, 1819—Waterdogs and Mudpuppies
    N. maculosus (Rafinesque, 1818)—Mudpuppy
        N. m. maculosus (Rafinesque, 1818)—Common Mudpuppy

Notophthalmus Rafinesque, 1820—Eastern Newts
    N. viridescens (Rafinesque, 1820)—Eastern Newt
        N. v. louisianensis Wolterstorff, 1914—Central Newt

Plethodon Tschudi, 1838—Woodland Salamanders
    P. cinereus (Green, 1818)—Eastern Red-backed Salamander

LIZARDS

Aspidoscelis Fitzinger, 1843—Whiptails
    A. sexlineata (Linnaeus, 1766)—Six-lined Racerunner
    A. s. viridis (Lowe, 1966)—Prairie Racerunner

Plestiodon Duméril and Bibron, 1839—Toothy Skinks
    P. fasciatus (Linnaeus, 1758)—Common Five-lined Skink
    P. septentrionalis (Baird, 1859 "1858")—Prairie Skink
        P. s. septentrionalis (Baird, 1859)—Northern Prairie Skink

SNAKES

Coluber Linnaeus, 1758—North American Racers, Coachwhips, and Whipsnakes
    C. constrictor Linnaeus, 1758—North American Racer
    C. c. foxii (Baird and Girard, 1853)—Blue Racer

Crotalus Linnaeus, 1758—Rattlesnakes
    C. horridus Linnaeus, 1758—Timber Rattlesnake

Diadophis Baird and Girard, 1853—Ring-necked Snakes
    D. punctatus (Linnaeus, 1766)—Ring-necked Snake
        D. p. arnyi Kennicott, 1859—Prairie Ring-necked Snake
        D. p. edwardsii (Merrem, 1820)—Northern Ring-necked Snake

Heterodon Latreille, 1801—North American Hog-nosed Snakes
    H. nasicus Baird and Girard, 1852—Plains Hog-nosed Snake
    H. platirhinos Latreille, 1801—Eastern Hog-nosed Snake

Lampropeltis Fitzinger, 1843—Kingsnakes
    L. triangulum (Lacépède, 1789)—Milksnake
        L. t. syspila (Cope, 1888)—Red Milksnake
        L. t. triangulum (Lacépède, 1789)—Eastern Milksnake

Nerodia Baird and Girard, 1853—North American Watersnakes
    N. sipedon (Linnaeus, 1758)—Common Watersnake
        N. s. sipedon (Linnaeus, 1758)—Northern Watersnake

Opheodrys Fitzinger, 1843—Greensnakes
    O. vernalis (Harlan, 1827)—Smooth Greensnake

Pantherophis Fitzinger, 1843—North American Ratsnakes
    P. obsoletus (Say, 1823)—Western Ratsnake
    P. ramspotti Crother, White, Savage, Eckstut, Graham and Gardner, 2011—Western Foxsnake

Pituophis Holbrook, 1842—Bullsnakes, Pinesnakes, and Gopher Snakes
    P. catenifer (Blainville, 1835)—Gophersnake
        P. c. sayi (Schlegel, 1937)—Bullsnake

Sistrurus Garman, 1884—Massasauga and Pygmy Rattlesnakes
    S. catenatus (Rafinesque, 1818)—Massasauga
        S. c. catenatus (Rafinesque, 1818)—Eastern Massasauga

Storeria Baird and Girard, 1853—North American Brownsnakes
    S. dekayi (Holbrook, 1836)—Dekay's Brownsnake
        S. d. texana Trapido, 1944—Texas Brownsnake
    S. occipitomaculata (Storer, 1839)—Red-bellied Snake
        S. o. occipitomaculata (Storer, 1839)—Northern Red-bellied Snake
        S. o. pahasapae Smith, 1963—Black Hills Red-bellied Snake

Thamnophis Fitzinger, 1843—North American Gartersnakes
    T. radix (Baird and Girard, 1853)—Plains Gartersnake
    T. sirtalis (Linnaeus, 1758)—Common Gartersnake
        T. s. parietalis (Say, 1823)—Red-sided Gartersnake
        T. s. sirtalis (Linnaeus, 1758)—Eastern Gartersnake

Tropidoclonion Cope, 1860—Lined Snakes
    T. lineatum (Hallowell, 1856)—Lined Snake

TURTLES

Apalone Rafinesque, 1832—North American Softshells
    A. mutica (Lesueur, 1827)—Smooth Softshell
        A. m. mutica (Lesueur, 1827)—Midland Smooth Softshell
    A. spinifera (Lesueur, 1827)—Spiny Softshell
        A. s. spinifera (Lesueur, 1827)—Eastern Spiny Softshell

Chelydra Schweigger, 1812—Snapping Turtles
    C. serpentina (Linnaeus, 1758)—Snapping Turtle

Chrysemys Gray, 1844—Painted Turtles
    C. picta (Schneider, 1783)—Painted Turtle
        C. p. bellii (Gray, 1831)—Western Painted Turtle

Emydoidea Gray, 1870—Blanding's Turtles
    E. blandingii (Holbrook, 1838)—Blanding's Turtle

Glyptemys Agassiz, 1857—Sculpted Turtles
    G. insculpta (LeConte 1830)—Wood Turtle

Graptemys Agassiz, 1857—Map Turtles
    G. geographica (LeSueur, 1817)—Northern Map Turtle
    G. ouachitensis Cagle, 1953—Southern Map Turtle
        G. o. ouachitensis Cagle, 1953—Ouachita Map Turtle
    G. pseudogeographica (Gray, 1831)—False Map Turtle
        G. p. pseudogeographica (Gray, 1831)—Northern False Map Turtle

Sternotherus Gray, 1825—Musk Turtles
    S. odoratus (Latreille, 1801)—Eastern Musk Turtle

Trachemys Agassiz, 1857—Sliders
    T. scripta (Schoepff, 1792)—Pond Slider
        T. s. elegans (Wied-Neuwied, 1838)—Red-eared Slider (introduced species)

SPECIES OF POSSIBLE OCCURRENCE

Found near the Minnesota border, but no records within the state:

Anaxyrus woodhousii (Girard, 1854)—Woodhouse's Toad
　　A. w. woodhousii Girard, 1854—Rocky Mountain Toad

Lithobates blairi (Mecham, Littlejohn, Oldham, Brown, and Brown, 1973)—Plains Leopard Frog

Spea bombifrons (Cope, 1863)—Plains Spadefoot

Class Amphibia

Class Amphibia

Amphibians evolved from fish ancestors around 400 million years ago. The evolution of lungs and limbs allowed ancestral amphibians to colonize land during the late Devonian period and become the first land-dwelling vertebrates. Modern-day amphibians consist of approximately 6,000 species in three orders. Restricted to the tropics, the Order Gymnophiona (caecilians) contains over 170 species of secretive, worm-like creatures. Members of the Order Caudata (salamanders) have four legs, a relatively long tail, and a head distinct from the body. There are approximately 400 species of salamanders worldwide, of which 100 are found in the United States. There are over 5,300 species of Order Anura (Salientia) (frogs and toads), of which 80 species are found in the United States. Frogs and toads are collectively referred to as anurans. They are tailless, lack a distinct neck, and have elongated rear legs adapted for jumping. Eight species of salamanders and 14 species of anurans are found in Minnesota.

The name *amphibian* means double life, referring to the aquatic and terrestrial phases found in the life cycles of many species. Most amphibians must live in water or in a relatively damp environment to prevent dehydration. Their glandular skin lacks an outer covering, such as scales or water-impermeable skin, to retard water loss. Amphibians' requirement for moisture also explains their greatly increased activity during or shortly after rainfall. By far, the largest diversity and concentrations of amphibians are found in regions with abundant rainfall.

All North American amphibians lay eggs. Lacking a protective shell, the eggs are generally laid in water. They hatch into gill-bearing larvae, which later transform into air-breathing adults. Two exceptions to this life cycle are found in Minnesota. Mudpuppies live an entirely aquatic life, and Eastern Red-backed Salamanders lack an aquatic larval stage.

Amphibians are the only class of land-dwelling vertebrates that have external gills during some phase of their life. Amphibian larvae always pass through a stage with external gills either

during development in the egg or later as a free-living animal. Nearly all amphibians lack claws; instead they have blunt toe tips or toe discs. Some species of toads have spurs on their rear feet that they use for digging.

Amphibian skin has several interesting features. Numerous mucus glands in the skin produce a slimy secretion that helps protect the skin, while the mucus itself is useful for slipping out of a predator's grip. Poison glands that are found in the skin of a number of species serve as another defense mechanism. The large parotoid glands behind the eyes of a toad consist of multiple poison glands. When handled roughly, toads release a whitish secretion from these glands that is extremely irritating to the membranes of the mouth of predators. Dogs have been fatally poisoned by chewing on certain species of tropical toads, which are not native to Minnesota.

Several species of treefrogs are capable of remarkable color changes. A hormone released from the pituitary gland causes a pigment shift in special skin cells called melanophores. Environmental temperature and the emotional state of the animal affect the color displayed. Males often display bright colors during courtship. A skin color that matches or blends with surroundings is useful for eluding predators. In Minnesota, the Gray Treefrog is capable of changing from dark gray to bright lime green.

Respiration is another important function of amphibian skin. Moist, highly vascular areas of their skin permit gaseous exchange with the environment. Salamanders of the family Plethodontidae are lungless and maintain a terrestrial existence by "breathing" through their skin and the lining of their mouth. The Eastern Red-backed Salamanders found in Minnesota are members of this family. Periodically, the amphibian's outer skin layer is shed and in many cases is consumed by its owner.

Courtship and reproduction are an important and interesting aspect of amphibian life history. Most species mass into breeding congregations to carry out courtship activities and egg laying. Salamanders normally concentrate in the spring or the fall. During these congregations males court females by pushing, rubbing, and nudging them. Fertilization is internal, although copulation does not occur. Following courtship, females use their cloacal lips to pick up gelatinous, sperm-filled sacs (spermatophores) deposited by males.

Frog and toad breeding season tends to be a raucous affair. Male anurans reach the breeding ponds first and begin vocalizing. Each species has its own distinctive voice, and several species may call from the same pond simultaneously. Noise produced in the male's larynx is amplified and resonated with single or paired vocal sacs. After females arrive, there is much pushing and shoving between males to establish dominance over an individual female. Male anurans clasp females behind their front legs in what is called amplexus, and the eggs are fertilized externally as she lays them. Some male frogs and toads are nonselective of mates and will mount other males. Mounted males may emit a guttural noise to signal the other male to release. Some species set up territories, which reduces fights over individual females and extends the length of the breeding season. The number of eggs laid may vary from a few hundred eggs to several thousand, depending on the species.

The delicate amphibian egg, which requires high moisture for survival and development, is covered with a gelatinous coat and is generally darkly pigmented. Dark pigment absorbs sunlight and increases egg temperature, expediting development of the embryo.

After a short incubation period, most species of amphibians exhibit a free-living aquatic stage that uses gills to obtain oxygen. Larval anurans are popularly called tadpoles or polliwogs. Their round body and flattened tail are familiar to almost everyone. Salamander larvae generally possess legs and external gills.

Amphibian transformation, or metamorphosis, is a fascinating and complex phenomenon. Rotund little tadpoles sprout legs, lose their tails, trade gills for lungs, develop a large mouth, and hop out onto land. Salamander larvae undergo a similar change but with fewer outward physical alterations. In Minnesota, transformation may take a few weeks or as long as two years, depending on the species. Most amphibians reach breeding age in 1 or 2 years. The average life span is 3 to 8 years. Some frog species are known to exceed 10 years, and Spotted Salamanders have exceeded 30 years (Wells 2007).

The ecological significance of amphibians is just becoming appreciated. Adult amphibians are entirely carnivorous predators. They have voracious appetites and consume extraordinarily large quantities of arthropods and other invertebrates. Not only do amphibians consume large quantities of animal protein, they in turn serve as an important food source for native fish, birds, reptiles, and mammals. Amphibians command an important position in the natural food web. The numbers of some amphibians, especially anurans, have declined sharply during the past decade. Herpetologists have expressed a deep concern about this decline and are trying to identify the causes of and find solutions to this compelling problem.

Family Bufonidae—True Toads

With over 550 species in the family, true toads are found on every continent except Antarctica, although the Cane Toads found in Australia were introduced. True toads are normally heavy bodied with wide heads and rough, dry-feeling skin, often covered with "warts." Their thicker epidermis slows moisture loss, allowing them to withstand relatively dry environments compared to true frogs and treefrogs. Other characteristics include parotoid glands behind the eyes, reduced webbing between the toes, and short legs for walking or hopping. The eggs of North American bufonids are laid in long gelatinous strings.

Minnesota's three representatives of this family, all in the genus *Anaxyrus,* are the American Toad, Great Plains Toad, and Canadian Toad.

American Toad

Anaxyrus americanus

DESCRIPTION

The American Toad is the most common toad in Minnesota. Body color is normally brown but can range from dark brown to tan and rarely is red or green. A white stripe may be present down its back. Variable white and black splotches occur on the back. The majority of the black spots on the back have one or two "warts" per spot. The belly is white with gray and black flecking, especially on the throat. The oval parotoid glands are separated from the cranial crest or attached by a narrow spur. Males have a single round vocal sac, which is darkened during the breeding season. American Toads' snout–vent length ranges from 5 to 9 centimeters (2 to 3½ inches) (Conant and Collins 1998). During the breeding season males have enlarged thumb pads for clasping onto females. Maturity is reached at three to four years of age for females (Acker, Kruse, and Krehbiel 1986; Kalb and Zug 1990), a year or two later than males (Hamilton 1934).

The warts on the femur are larger than those found on the tibia. This characteristic distinguishes the American Toad from

American Toad, adult. Photograph by Allen Blake Sheldon.

the Woodhouse's Toad, which has similar-sized warts on the femur and tibia. The parotoid glands and cranial crests are normally fused in the Woodhouse's Toad, but this characteristic is variable.

Toad tadpoles are small and dark colored. They reach a length of 2.5 centimeters (1 inch). The body color is gray black with clear tail fins. The fins do not extend onto the body and are rounded at the tip. The three species of toads in Minnesota have very similar-looking tadpoles and therefore are impossible to identify without very close examination.

Distribution

American Toads are found east of the Great Plains and north of the Gulf of Mexico coastal plain. The northern limit is Hudson Bay in Quebec and Ontario. In Minnesota, the American Toad is found statewide. The western border of the state is the approximate western edge of the toad's national range.

HABITAT

American Toads are found in a wide variety of habitats in Minnesota and are quite tolerant of habitat fragmentation (Lehtinen, Galatowitsch, and Tester 1999). They are found in the bogs and coniferous forests of the north (Karns 1984, 1992a) and throughout wooded areas of central and southern Minnesota. Prairies provide good habitat in the southwest part of the state, but the prairies of the Red River valley lack American Toads. Ewert (1969) described the American Toad as a woodland toad, and the Great Plains Toad as a prairie toad. This may be true on the northwestern Minnesota prairies, but American Toads are regularly found on Minnesota's west-central and southwestern prairies (Moriarty 1988).

Calling male American Toad. Photograph by Tony Gamble.

American Toads utilize a variety of aquatic habitats for breeding including shallow wetlands, ditches, and backwaters of rivers and streams.

LIFE HISTORY

American Toads overwinter by burrowing into well-drained, preferably sandy soils. Winter dormancy normally begins in early October, but toads have been seen active as late as 24 October. They burrow below the frost line (Ewert 1969; Tester, Parker, and Siniff 1965), which varies in depth from approximately 12 to 58 centimeters (5 to 23 inches). Burrow depths are adjusted as the frost line shifts. When conditions are extremely dry during the summer, shallow burrows are used for estivation.

American Toads are more cold tolerant than other toad species in Minnesota, emerging from winter dormancy by late April. Movements to breeding pools occur shortly after emergence. Breeding normally occurs from early May to mid-June. Toads have a moderate fidelity to breeding sites (Ewert 1969), which include ponds, lakes, rivers, and swamps. Breeding congregations are normally large, and dozens of amplexing pairs of toads can commonly be seen floating near the edge of ponds. During the breeding season male toads attempt to amplex with anything that swims by, including other male toads or frogs.

After breeding they have been recorded moving as far as 1,000 meters (3,300 feet), but most move much shorter distances (Ewert 1969). These movements are nocturnal, and activity peaks on warm, wet evenings. American Toads tend to rest at the ground's surface or under leaf litter during the day. In the fall, movements back to overwintering areas tend to be diurnal because of cool temperatures in the evenings.

The male's call is a high-pitched, rapid trill that lasts 20 to 30 seconds. The trill is a single note with little variation in pitch. Calling begins at dusk but may continue throughout the day if it is cloudy or raining.

Females mature in their third year and lay 4,000 to 20,000 eggs (Collins 1982). Eggs hatch in two to eight days depending on the water temperature. Warmer temperatures speed up the development of the eggs. Tadpoles school and look like a large black mass moving around the shallows of ponds or streams. They generally

American Toads in amplexus.
Photograph by Jeffrey B. LeClere.

American Toad tadpoles. Photograph by Allen Blake Sheldon.

emerge as toadlets between late June and July, but they are capable of speeding up metamorphosis if trapped in drying pools. Toadlets are less than 1 centimeter (½ inch) in length when they transform. During years with high production, large numbers of small toadlets emerge from wetlands and cover the ground as they disperse across the landscape. Homeowners often see them all over their yards in July and August when mowing the lawn.

American Toads feed on a variety of terrestrial insects and invertebrates. Feeding activity is nocturnal except during rainy weather. Toads stalk their prey for short distances. They use their tongues to capture prey, and then they use their front legs to stuff the prey into their mouth. American Toads are eaten by a variety of birds, but their primary predators are snakes, especially hog-nosed snakes that feed almost exclusively on toads.

Tadpoles have toxins in their skin, making them unappealing to many potential predators (Petranka and Hayes 1998). Aquatic invertebrates with sucking mouthparts, such as the diving water beetle, avoid these toxins by piercing through the tadpole's skin and sucking out the bodily fluid.

Adult American Toads defend themselves by inflating their bodies to look larger and to prevent being swallowed. If the bluff fails and the attack continues, the toads tuck their heads down and present their parotoid glands. The glands secrete a toxin that can make their attacker sick (DeGraaf 1991). Toads have also been known to urinate on their attacker. This seems to be the best defense when picked up by humans!

Two mammal species that prey on adult toads are raccoons and striped skunks, both of which will eviscerate the toads, leaving the skin and remaining carcass (Groves 1980; Schaaf and Garton 1970), thus avoiding the toxins.

REMARKS

The Eastern American Toad (*A. a. americanus*) is the subspecies found in Minnesota (Conant and Collins 1998).

In Oldfield and Moriarty (1994), the Eastern American Toad was in the genus *Bufo*. Frost et al. (2006) found the genus to have several evolutionary lineages and removed North American toads from *Bufo* and put them into the new genus *Anaxyrus*.

The common myth of toads transmitting warts to humans has no factual basis. Warts in humans are caused by a virus. The "warts" on toads are skin glands.

An albino American Toad was collected in Le Sueur County in 1992. This is the first record of an albino American Toad from Minnesota.

Hybridization is known to occur with Canadian Toads in South Dakota and Canada (F. R. Cook 1983; Henrich 1968; Green and Pustowka 1997). This appears to be common in western and southwestern Minnesota as well, where individual toads have been observed with features of both species. Breeding calls of hybrid American-Canadian Toads are shorter with a deeper tone than American Toads, often lasting only five to seven seconds.

Although populations appear secure statewide, low pH levels are lethal to American Toad larvae (Tattersall and Wright 1996), and chemical contaminants, including pesticides, can result in malformations. Limb and eye malformations are some of the reported cases in Minnesota. Flesh-eating fly larvae have been observed parasitizing toads in Minnesota. Fly parasitism could be a major source of mortality in a population, but this has not been documented (Wells 2007). In some studies frogs survive the infection with little obvious tissue damage; others may die at the time of larval emergence (Elkan 1965).

Great Plains Toad

Anaxyrus cognatus

DESCRIPTION

The Great Plains Toad is a grassland species tolerant of the dry prairie environment. It can be identified by its distinctive large, dark spots with light edges on its back, and each spot has several small warts. Background color may be gray brown to green. The belly is a uniform white or ivory with no black flecking. Adult males have a dark throat (Bragg 1940) with an oblong, sausage-shaped vocal sac that extends beyond the snout when inflated. The oval parotoid glands are not as prominent as those of the American or Canadian Toads. The cranial crests are reduced and form a V between the eyes. Great Plains Toads' snout–vent length ranges from 5 to 9 centimeters (2 to 3½ inches), which is similar in size to the American Toad. Their hind feet often have sharp-edged tubercles for digging. Females are larger than males (Wright and Wright 1949).

Great Plains Toad tadpoles are very similar in appearance to American Toad tadpoles (see the American Toad species account for a detailed description).

Great Plains Toad, adult. Photograph by Allen Blake Sheldon.

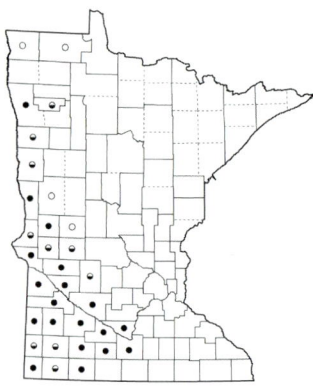

DISTRIBUTION

The distribution of the Great Plains Toad is similar to its common name. It is found in the Great Plains of North America from Montana to Minnesota, and south into Mexico and the Colorado River drainage. It has been described as expanding eastward into northwestern Iowa (Lannoo et al. 1994). In Minnesota the species is found along the western border of the state from Marshall County south to Rock County. The eastern edge of the range is in Nicollet County.

Great Plains Toad breeding habitat in western Minnesota. Photograph by Minnesota Department of Natural Resources—Carol D. Hall.

HABITAT

Great Plains Toads inhabit open landscapes such as grasslands, floodplains, and agricultural fields. They are often associated with well-drained soils and cultivated fields in stream valleys. In Minnesota they use tallgrass prairies and nonnative grasslands. They are rarely found in woodlands (Ewert 1969). The toads select high, dry ground for overwintering sites. They burrow deeper than American Toads, up to 100 centimeters (40 inches) (Ewert 1969).

Adult male Great Plains Toad calling from flooded cornfield. Photograph by Minnesota Department of Natural Resources—Carol D. Hall.

LIFE HISTORY

The Great Plains Toad is the most fossorial of Minnesota's toads and is a very rapid burrower (Ewert 1969). This characteristic is an adaptation to living in dry grassland habitats. Emergence and breeding are triggered by early summer thunderstorms between mid-May and mid-July (Breckenridge 1944). During years of inadequate rainfall, reproduction may not occur (Sullivan and Fernandez 1999). With adequate rainfall, the toads emerge and move immediately to the shallow breeding pools. Bragg (1940) found buffalo wallows were readily used as breeding sites. In Minnesota, Great Plains Toads often breed in shallow wetlands that form in agricultural fields after heavy rain events. Females may lay over 40,000 eggs and can lay multiple clutches per breeding season (Krupa 1986, 1994). Eggs typically hatch in two to seven days (Krupa 1994). The period from fertilized egg to metamorphosis is temperature dependent and ranges from less than three weeks (Hahn 1968; Krupa 1994) to six weeks (Breckenridge 1944). Young may aggregate in large numbers after metamorphosis (Bragg and Brooks 1958).

After breeding, their movements to foraging areas range from 300 to 1,300 meters (1,000 to 4,000 feet). Daily movements are approximately 30 meters (100 feet) (Ewert 1969). Great Plains Toads estivate in the summer during dry, hot weather. If weather conditions are not favorable, this toad commonly remains underground through winter without coming out of estivation.

The mating call of the Great Plains Toad is harsh with a pulsating mechanical quality, in contrast to the steady musical trill of the American Toad (Breckenridge 1944). Individual calls last from 20 to 50 seconds (Conant and Collins 1998).

Great Plains Toads feed on terrestrial insects. Beetles and ants have been reported as prey items (Collins 1982). Hammerson (1982) reports that this toad is an effective predator on agricultural pests such as cutworms, moths, and flies. The predators of Great Plains Toads are similar to those of the American Toad.

REMARKS

In Oldfield and Moriarty (1994), the Great Plains Toad was in the genus *Bufo*. Frost et al. (2006) found the genus to have several evolutionary lineages and removed North American toads from *Bufo* and put them into the new genus *Anaxyrus*.

Breckenridge (1944) reported that no toads were observed in the western prairies in the early 1930s, but when it rained in 1937, the Great Plains Toads emerged in great numbers. Great Plains Toads were not found on the prairies during 1988 (Moriarty 1988), which was the second year of a drought.

The status of Great Plains Toads in Minnesota is uncertain. Although local breeding populations are evident during rain events, recruitment of young has likely been reduced over time. Wetland drainage has undoubtedly taken a toll on this species, which requires at least a three-week hydroperiod for breeding and development of young. Populations are difficult to monitor due to this toad's fossorial nature and unpredictable breeding season.

The Great Plains Toad is listed as special concern in Minnesota (Minnesota DNR 2013).

Canadian Toad

Anaxyrus hemiophrys

DESCRIPTION

This species resides in the prairies and aspen parklands, often near water. The distinguishing feature of the Canadian Toad is a large boss between the eyes. The boss is a raised bump formed by the fusion of the cranial crests. The body color is different shades of brown, with variable patterns of white and black. The dominant spots are black, with one or two warts per spot. The belly is white with gray and black flecking, especially on the throat. Males have a single round vocal sac. Canadian Toads' snout–vent length ranges in size from 5 to 9 centimeters (2 to 3½ inches) (Conant and Collins 1998). When Canadian Toads first transform, they are less than 1 centimeter (½ inch) long from snout to vent and look identical to American Toads. The cranial boss is the only characteristic that separates the appearance of these two species.

Canadian Toad tadpoles are very similar in appearance to American Toad tadpoles (see the American Toad species account for a detailed description).

Canadian Toad, adult.
Photograph by Jeffrey B. LeClere.

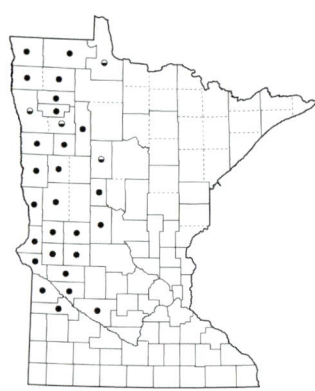

DISTRIBUTION

The range of the Canadian Toad in the United States is northwest Minnesota and North Dakota. In Canada they are found on the prairies of Manitoba and Saskatchewan. In Minnesota they are found from the Minnesota River in Yellow Medicine and Renville Counties north through the Red River valley.

HABITAT

Canadian Toads are found in the same habitats as American Toads, ranging from woodlands to wetlands. They are not found in the interior of vast expanses of prairie (Ewert 1969). They overwinter on upland sites approximately 25 meters (80 feet) from wetlands. Breeding habitat includes ditches, ponds, and shallow bays of lakes.

During the summer, adult toads are found close to water. They are the most aquatic of Minnesota's toads (Breckenridge and Tester 1961).

LIFE HISTORY

The overwintering habits of Canadian Toads differ from American Toads'. Canadian Toads hibernate communally, whereas American Toads hibernate individually. They congregate in Mima mounds, which are small earth mounds 3 to 12 meters (9 to 39 feet) in diameter and 1 meter (3 feet) high (Breckenridge and Tester 1961). The name *Mima mounds* was derived from the Mima Prairie of Washington, characterized by its numerous small hillocks. These mounds, thought to be created by gopher activity, are comprised of coarse weedy vegetation and looser soils compared to the surrounding prairie. Tester and Breckenridge (1964) found that 99 percent of the toads were in

Close-up of Canadian Toad. Photograph by James E. Gerholdt.

Mima mounds, with an average of 950 toads per mound, whereas less than 1 percent of the toads were in the surrounding uplands. Toads typically return to the same Mima mound (Kelleher and Tester 1969). Like other wintering toads, they burrow during the winter to stay just below the frost line (Tester 1981; Tester and Breckenridge 1964).

The breeding habits of Canadian Toads are similar to those of American Toads. Canadian Toads emerge as the ground thaws in the spring. Males exit first and move to breeding sites in late April or early May (Tester and Breckenridge 1964). Males have a call similar to American Toads', except that the trill length is much shorter (2 to 8.5 seconds; Preston 1982). Females lay long strings of eggs with 3,000 to over 5,000 eggs laid per female (Roberts and Lewin 1979). Eggs hatch in several days. The tadpoles metamorphose in late June and early July, requiring roughly six to eight weeks for development (Breckenridge 1944). They emerge as tiny toadlets between 9 and 15 millimeters (⅓ to ⅔ inch) in length (Wright and Wright 1949; Breckenridge and Tester 1961).

Daily movements can range from nearly no movements to more than 225 meters (250 yards) (Breckenridge and Tester 1961), similar to the range of American Toads.

REMARKS

In Oldfield and Moriarty (1994), the Canadian Toad was in the genus *Bufo*. Frost et al. (2006) found the genus to have several evolutionary lineages and removed North American toads from *Bufo* and put them into the new genus *Anaxyrus*.

Some researchers (F. R. Cook 1983; Preston 1982) feel that Canadian Toads and American Toads are subspecies because of the presence of an intergradation zone in Manitoba and Ontario, Canada, where successful breeding occurs between the two species. Green (1983), however, treats them as separate species that sometimes hybridize when their ranges overlap. They are known to breed with American Toads in South Dakota (Henrich 1968), and hybrids were common during surveys in 2006 and 2007 in southwest Minnesota.

Family Hylidae—Treefrogs

Treefrogs are found on every continent except Antarctica. This family of frogs is one of the largest, with over 800 species. Treefrogs are slender bodied, with a flattened profile. Most have discs on their toes (toe pads) to assist in climbing. The majority of species are arboreal, but some species have evolved to live in terrestrial or semiaquatic habitats.

Minnesota has five species of hylids in three genera: Blanchard's Cricket Frog, Cope's Gray Treefrog, Gray Treefrog, Spring Peeper, and Boreal Chorus Frog. The two Gray Treefrogs are the only highly arboreal frogs in Minnesota.

Blanchard's Cricket Frog
Acris blanchardi

DESCRIPTION

Blanchard's Cricket Frogs are small frogs with a snout–vent length ranging from 2 to 3.8 centimeters (⅝ to 1½ inches). The skin color is brown ranging to gray, with variable green blotches. Darker-colored warts are present on the dorsal surface. Most Blanchard's Cricket Frogs have a green or dark-brown triangle between the eyes pointing backwards, and they may have a light middorsal stripe. There are dark bars on the upper jaw. They lack the enlarged toe pads present on their arboreal relatives. Hind feet are webbed, affording this semiaquatic treefrog an advantage in aquatic habitats. The sexes are similar, although females are slightly larger than males. Also, males develop a dark vocal sac during the breeding season.

Blanchard's Cricket Frog tadpoles are the only tadpoles in Minnesota with a black-tipped tail. However, larvae developing where fish are present may lack the black coloration (J. P. Caldwell 1982). Tadpoles reach a length of 3.5 centimeters (1½ inches).

Blanchard's Cricket Frog, adult. Photograph by Allen Blake Sheldon.

DISTRIBUTION

Blanchard's Cricket Frogs are found in the central and southern United States. Historically they were found from southern Ontario south to the Ohio River and west to Nebraska and Texas. Although its southern range is primarily west of the Mississippi River, populations also exist in Mississippi and Kentucky.

Formerly considered a common species throughout much of its range, declines were first noted in the 1970s and 1980s (P. W. Smith 1961; Minton 1972; Campbell 1977; Vogt 1981). Since then, Blanchard's Cricket Frog populations have declined throughout the northern portion of its range (Gray, Brown, and Blackburn 2005) and are considered extirpated from Ontario.

Blanchard's Cricket Frogs have been recorded from the southeastern and southwestern corners of Minnesota, and an outlying record exists from Chisago County. The only recent records are from Hennepin, Wabasha, and Winona Counties.

HABITAT

Blanchard's Cricket Frogs are considered terrestrial and semi-aquatic, remaining close to permanent water bodies including lakes, ponds, streams, and rivers. They forage and loaf along muddy shorelines with abundant emergent vegetation and openings of mud or gravel (Smith et al. 2003; Russell et al. 2002).

Blanchard's Cricket Frogs escape freezing temperatures by taking refuge underground and may hibernate communally (McCallum and Trauth 2006). In Ohio they were found hibernating up to 30 centimeters (12 inches) below the soil surface adjacent to streams and ponds (Walker 1946). In North Carolina cricket frogs overwinter in rock crevices and tree roots 549 meters (1,800 feet) from their summer habitat (Westerveld 2012). The northern range of this species may be limited by the availability of suitable hibernacula (Swanson and Burdick 2010).

LIFE HISTORY

In Minnesota, Blanchard's Cricket Frogs emerge from winter dormancy in late April. They breed from late May into July along with Green Frogs and American Bullfrogs. Breeding choruses are stimulated by rain and can be heard in the day as well as at

Calling male Blanchard's Cricket Frog. Photograph by Allen Blake Sheldon.

night. Males call from the water or along the shoreline. Their call is an insect-like "glick, glick, glick" sound similar to the clicking of two rocks or metal balls. Identification by call can be difficult because Virginia Rails, which are found in many of the same habitats, make a similar sound. Blanchard's Cricket Frogs normally call in choruses of more than one male, which may help distinguish the call from the usually solitary rail's. Females lay up to 400 eggs (Livezey 1950) in clumps of 10 to 15 eggs each. In Wisconsin, Vogt (1981) found clumps of eggs attached to vegetation in flowing water.

Tadpoles hatch within a few days after the eggs are laid. Five to 10 weeks after hatching, the tadpoles undergo metamorphosis. Young frogs grow rapidly and stay active into late September, whereas adults become inactive in August (Johnson and Christiansen 1976). These small frogs feed on tiny insects, eating enough to fill their stomachs three times a day (Johnson and Christiansen 1976). Frogs mature within their first year. Young toads can easily be mistaken for Blanchard's Cricket Frogs.

Blanchard's Cricket Frogs stay near water throughout the summer (Vogt 1981) but may disperse to distant aquatic sites after rain events (Gray 1983; Burkett 1984). They escape predators by using their remarkable jumping skills. A cricket frog can jump up to 1 meter (3 feet) in a single leap (Vogt 1981). They will leap into the water, then immediately return to the shoreline, swimming near the surface of the water. During periods of drought they may seek refuge in cracks of dry wetland basins (Blair 1957).

REMARKS

In Oldfield and Moriarty (1994), Blanchard's Cricket Frog was referred to as the Northern Cricket Frog (*Acris crepitans blanchardi*). Gamble et al. (2008) found *Acris blanchardi* to be a distinct species based on molecular evidence, elevating it to species level.

Blanchard's Cricket Frog (*Acris blanchardi*) is named in honor of herpetologist Frank Nelson Blanchard (1888–1937) (Harper 1947), a professor at the University of Michigan.

Surveys conducted in Minnesota in the early and mid-1990s were unsuccessful in documenting populations in its former range (Hall 1997; Van Gorp 1996; Van Gorp and VanDeWalle 1995). In 1998, a Blanchard's Cricket Frog population was discovered along the Minnesota River in Hennepin County (Moriarty, Forbes, and Jones 1998), and additional discoveries were made in Winona County in 2004 and Wabasha County in 2012. The origins of these populations are suspect, as they may have been introduced or transported via a barge or flood. However,

DNA analysis confirmed that the population along the Minnesota River was similar to the genetic makeup of *Acris blanchardi* from surrounding states and could be an isolated remnant population (Berendzen, Gamble, and Simons 2003). Results of call surveys for these isolated populations indicate that Blanchard's Cricket Frogs are persisting but remain vulnerable to catastrophic events.

The decline of Blanchard's Cricket Frogs from the northern part of their North American range is likely due to multiple factors including habitat fragmentation, pesticide use, climate change, and catastrophic events. In addition to these factors, this species has an unusually short life span, which compromises its ability to recover from repeated years of high mortality and low recruitment.

Burkett (1984) found extremely low winter survival of cricket frogs in a Kansas population, in which only 5 percent survived to breed the following year. The life history of this species appears to be dependent on high recruitment of young into the population, limiting the potential of the species to recover from consecutive annual catastrophic events.

The Blanchard's Cricket Frog is listed as endangered in Minnesota and is considered a Species of Greatest Conservation Need by the Minnesota DNR (2006, 2013).

Cope's Gray Treefrog

Hyla chrysoscelis

DESCRIPTION

The Cope's Gray Treefrog is a medium- to large-sized treefrog with a snout–vent length ranging from 3 to 5 centimeters (1¼ to 2 inches) (Conant and Collins 1998). The toe pads are well developed and easily visible. The slightly granular skin changes color from mottled gray to green depending on temperature and habitat. Green coloration normally occurs in warmer temperatures and when the treefrogs are on green vegetation. During the breeding season adults tend to be green without any blotches. When the frogs are gray, large dark blotches are visible on the back. The insides of the thighs and shanks are yellow. In Wisconsin, Jaslow and Vogt (1977) found the average snout–

vent length of Gray Treefrogs *(Hyla versicolor)* to be slightly larger than that of Cope's Gray Treefrogs, but this is not a reliable distinguishing feature. The two frogs were thought to be the same species until 1968 (Ralin 1968). Cope's Gray and Gray Treefrogs are almost morphologically identical and can only be reliably separated by their calls and chromosome analysis. Mature males are smaller than mature females and have a dark vocal sac.

Tadpoles are high finned, with the top fin extending onto the upper body. The tail is mottled with black. The fin may have an orange or red tinge. A mature tadpole will reach 3.5 centimeters (1½ inches).

Cope's Gray Treefrog, adult. Photograph by Allen Blake Sheldon.

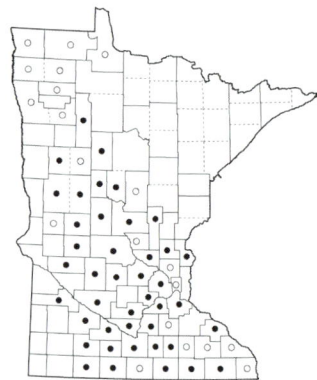

DISTRIBUTION

The overall distribution of the Cope's Gray Treefrog is thought to be the same as that of the Gray Treefrog. They range east of the Great Plains from a line through eastern North Dakota to central Texas east to the Atlantic coast.

In Minnesota, Cope's Gray Treefrogs occur in the prairie-forest transition zone that runs diagonally through the state. They are absent from the heavily forested counties of northeastern Minnesota, and records are sparse in many of the far western counties, particularly the southwest. Records prior to 1968 may be included with those of the Gray Treefrog.

HABITAT

Cope's Gray Treefrogs are associated with prairie edges and oak savannas (Vogt 1981). They are not found in forest interiors but do frequent woodland edges, where they overlap with Gray Treefrogs. They breed in shallow emergent marsh and shrub swamp habitats.

Cope's Gray Treefrog in amplexus. Photograph by Allen Blake Sheldon.

Calling male Cope's Gray Treefrog. Photograph by Allen Blake Sheldon.

LIFE HISTORY

The life history of the Cope's Gray Treefrog is similar to the Gray Treefrog's. Both species spend their winter in a partially frozen state under leaf litter, rocks, or logs (Schmid 1982). Glycerol is produced, acting as antifreeze to prevent vital organs from freezing (Irwin and Lee 2003). Frogs from Minnesota tend to store more glycogen in the liver, possibly as an adaptation to prolonged periods of cold weather (Irwin and Lee 2003). After emerging in mid- to late April, they spend time feeding in the uplands. They migrate to breeding ponds in mid-May and breed through June.

At the breeding ponds, males call from surrounding trees and emergent vegetation during the day or early evening. The calling perches of Cope's Gray Treefrogs range up to 3 meters (10 feet) above pond level in trees (Vogt 1981). The male's call is a fast metallic trill, compared to the Gray Treefrog's slower, more musical trill. The speed of the trill can vary with the temperature, faster when it is warm and slower when it is cool. Cope's Gray Treefrogs breed in deeper water than chorus frogs (Walker 1946). They lay up to 2,000 eggs singly or in small clusters. The tadpoles transform in 8 to 10 weeks.

Young frogs leave the pond in mid-July to join the frogs feeding in the shrubs and trees. During dry periods, both juveniles and adults estivate in hollow logs.

Cope's Gray Treefrog, tadpole.
Photograph by Allen Blake
Sheldon.

REMARKS

H. chrysoscelis is a normal diploid, having two sets of chromosomes (24 total), while *Hyla versicolor* is a tetraploid with four sets (48 total). Analysis of mitochondrial DNA has found that *H. versicolor* evolved from at least three different *H. chrysoscelis* lineages, ultimately forming one unique tetraploid species (Holloway et al. 2006). Naturally occurring hybrid triploids (*Hyla chrysoscelis* x *versicolor*) have also been documented (Gerhardt et al. 1994) and have a breeding call intermediate between their two parental forms. This rare occurrence is selected against under natural conditions due to the specific breeding call preferences of females.

A "Blue" Cope's Gray Treefrog was documented in Wright County in 1990. In 2009 and 2012 blue treefrogs were also observed in Hennepin County, although the species was not determined. These frogs were normal in every respect except that they were blue rather than green. This occurs when there is an absence of yellow pigment in the skin and is referred to as axanthism.

Antipredator mechanisms are present in tadpole and adult life stages. Colorful pigments in the tadpole's large tail appear to misdirect strikes by predators (e.g., dragonfly larvae or crayfish), protecting the palatable tadpole body (Kats, Petranka, and Sih 1988; Figiel and Semlitsch 1991; Van Buskirk et al. 2003). In adults, the flash of yellow on their inner thighs acts as a warning to potential predators. Adults may secrete skin toxins as a defense mechanism, which causes irritation to mucous membranes (e.g., eyes) of potential predators.

See also "Remarks" under Gray Treefrog.

Gray Treefrog
Hyla versicolor

DESCRIPTION

The Gray Treefrog is a medium- to large-sized treefrog from 3 to 5 centimeters (1¼ to 2 inches) in snout–vent length (Conant and Collins 1998). Their granular skin is gray with flecking and irregular, dark blotches, and the blotches have black borders. The skin color can change from gray to green, except for the black borders, which always remain visible in adults. Like the Cope's Gray Treefrog's, the inner surfaces of the thighs and shanks are bright yellow. Gray Treefrogs' large toe pads enable them to be very good climbers. They have rough skin, whereas Cope's Gray Treefrogs have smooth skin.

Gray Treefrog tadpoles are very similar in appearance to Cope's Gray Treefrog tadpoles. Tadpoles of both species are high finned, with the top fin extending onto the upper body. The tail is mottled with black. The fin may have an orange or red tinge. A mature tadpole will reach 3.5 centimeters (1½ inches).

Gray Treefrog, adult. Photograph by Barney Oldfield.

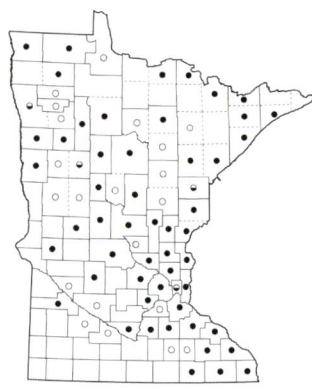

DISTRIBUTION

Gray Treefrogs are found from the Great Plains east to the Atlantic coast, except for northern Maine and southern Florida. In Minnesota they are found statewide except for the southwest corner.

HABITAT

Gray Treefrogs are found in a wide range of woodland habitats, both deciduous and coniferous. They use marshes and bogs associated with woodlands. They are not found on prairies but use floodplain forests in the prairie region.

Gray Treefrog breeding habitat. Photograph by Minnesota Department of Natural Resources—Kelly Lynch Pharis.

LIFE HISTORY

Gray Treefrogs overwinter under leaf litter, rocks, or logs in a manner similar to Cope's Gray Treefrogs. Their unusual overwintering strategy was first documented in Minnesota (Schmid 1982). Glycerol is accumulated in their body fluids during the fall when the temperature begins to cool. This protects the cells from rupturing when the frog partially freezes. When the frog begins to freeze, the liver converts the glycerol into glucose. Glucose is then circulated to major organs so that ice crystals will not form in the organ tissues. Ice forms in the body cavity around the organs and in between muscle cells. Up to 65 percent of the total body water is frozen (Storey and Storey 1990). Schmid (1982) found that the frogs can be cooled down to −7.2°C (19°F) for weeks and survive.

Gray Treefrog, metamorph. Photograph by Allen Blake Sheldon.

Axanthic Gray Treefrog. Photograph by Tony Gamble.

After emerging in the spring, Gray Treefrogs spend time feeding in the uplands. They move to breeding ponds in mid-May and breed through June. Males call from perches as high as 10 meters (30 feet) up in surrounding trees during the day or early evening (Vogt 1981). Any treefrog calling from above 3 meters (10 feet) is a Gray Treefrog because the Cope's Gray Treefrog does not call from above that height (Vogt 1981). The call, a melodic trill, is slower than the Cope's fast metallic trill. The trill rate is affected by the ambient temperature at the breeding pond. The trill rate is faster in warmer temperatures and similar to the call of the Cope's Gray Treefrog from cooler temperatures, but the trill rates of the two species do not overlap at a given temperature (Jaslow and Vogt 1977). Females lay up to 2,000 eggs singly or in small clusters. The tadpoles transform in 8 to 10 weeks.

Young frogs leave the pond to join the adults in late July or early August, feeding in the shrubs and trees, sometimes up to 10 meters (30 feet) above the ground. Gray Treefrogs feed on a variety of insects, but beetles and caterpillars comprise the majority of the diet (Vogt 1981). During the day, Gray Treefrogs spend their time under loose bark, in tree cavities, or in bird nest boxes (McComb and Noble 1981).

REMARKS

In Missouri, Johnson et al. (2007) found that Gray Treefrogs move 200 meters (219 yards) or more between foraging areas, overwintering sites, and breeding ponds; the maximum movement recorded is 330 meters (361 yards). Females tend to travel farther from breeding ponds than males.

Before 1968 (Ralin 1968), the Gray Treefrog and Cope's Gray Treefrog were considered to be the same species. The Gray Treefrog is a tetraploid (twice the normal number of pairs of chromosomes), and Cope's Gray Treefrog is a diploid (two pairs of each chromosome). Cope's Gray Treefrog has 24 chromosomes, and the Gray Treefrog has 48. Thus, absolute identification can only be made by chromosome analysis. Holloway et al. (2006) studied the complex evolution of Gray Treefrogs and found that they are composed of a single interbreeding lineage that was created from at least three distinct diploid species (Cope's Gray Treefrog, Bird-voiced Treefrog (*H. avivoca*), and an unknown species). This unusual polyploid formation had never been seen before.

A long history of taxonomic confusion makes it difficult to analyze distribution and habitat information prior to 1968. Early publications (Breckenridge 1944; Wright and Wright 1949) combined information on Minnesota's two species of Gray Treefrogs, so the life history differences are difficult to determine. Jaslow and Vogt (1977) demonstrated that Cope's Gray Treefrog is a prairie-associated species and the Gray Treefrog is a forest-associated species.

Spring Peeper
Pseudacris crucifer

DESCRIPTION

Spring Peepers are among the smallest frogs in Minnesota. They range from 1.9 to 3.2 centimeters (¾ to 1¼ inches) in snout–vent length. Skin color varies from light brown to gray to dark brown. Individuals are sometimes a bright-bronze color. Temperature affects the color of the frog. Colder frogs are dark, and warmer frogs are light in color. The Spring Peeper has a distinctive X on its back, thus its species name crucifer, which is Latin for cross-bearer. The belly is light tan with no markings. The toe pads are reduced but distinct.

Spring Peeper tadpoles are small, averaging a total length of 3 centimeters (1⅛ inches). Tail fins are medium in height and are heavily mottled but have a clear band adjacent to the tail muscle.

Spring Peeper, adult. Photograph by Allen Blake Sheldon.

DISTRIBUTION

Spring Peepers are found throughout the eastern United States east of the Great Plains and south of James Bay, Canada (51°N latitude). In Minnesota, Spring Peepers are found in the non-prairie counties.

HABITAT

Spring Peepers are normally found in woodlands, both deciduous and coniferous, where they spend most of their time near wetlands. Suitable breeding ponds are associated with upland wooded habitat and support emergent vegetation but are too shallow and temporary to contain fishes (Minton 2001).

Spring Peepers are usually absent from urban and agricultural areas because the uplands surrounding wetlands have been intensively developed.

LIFE HISTORY

Spring Peepers normally emerge in early April, about the time the ground thaws. They overwinter on land under logs or leaf litter. As in the bodies of the Gray Treefrog and Wood Frog, the production of glucose limits cellular freezing to 65 percent of the Spring Peeper's body during winter months (Schmid 1982; Churchill and Storey 1996).

Spring Peepers start breeding in April, soon after emergence, and continue through May. They typically initiate calling near the end of the Wood Frog's breeding season. Their call is a high-pitched peep or ping, which can be heard emanating from low vegetation in wetlands. Large choruses sound like sleigh bells and can be deafening when close. Eggs are laid singly (Walker 1946) or in small clusters that are attached to submerged vegetation; up to 1,000 eggs are produced in a season. Tadpoles take 12 to 14 weeks to transform after hatching (Wright 1914).

Spring Peepers feed on small insects and invertebrates, as do Minnesota's other small frogs. Most feeding is done while hunting in low vegetation (McAlister 1963).

Calling male Spring Peeper. Photograph by Jeffrey B. LeClere.

Spring Peeper egg mass. Photograph by Allen Blake Sheldon.

REMARKS

Annual frog and toad call surveys indicate a slight decline in Spring Peeper populations based on statewide analysis of 12 years of data (Larson 2009). Within the Twin Cities metropolitan area they have all but disappeared due to urbanization, although populations persist in outlying sites with suitable habitat.

The Spring Peeper was in the genus *Hyla* until 1986 (Hedges 1986).

Spring Peeper, tadpole. Photograph by Allen Blake Sheldon.

Boreal Chorus Frog
Pseudacris maculata

DESCRIPTION

Boreal Chorus Frogs are the most common treefrog in Minnesota, though they are rarely seen due to their small size. They have a maximum snout–vent length of 3.2 centimeters (1¼ inches), similar in size to the Spring Peeper. Females are slightly larger than males. Their body shape is slender with a pointed head. Skin color is variable and ranges from light brown to shades of red, gray, or green. The Boreal Chorus Frog has three, sometimes broken, longitudinal stripes on its back. The belly is tan with no markings. Color morphs of these frogs have been studied by Hoppe (1981). He reported that the brown phase is dominant in the warmer prairie areas of Minnesota and that a higher proportion of red and green morphs are found in the forested, cooler parts of the state. Toe pads are greatly reduced and are difficult to see with the naked eye. The toes are partially webbed.

Boreal Chorus Frog tadpoles lack the mottling found in other hylid tadpoles. They are small, with a maximum length of 3 centimeters (1⅛ inches). The tail fins are clear with the dorsal fin extending onto the body.

Boreal Chorus Frog, adult. Photograph by James E. Gerholdt.

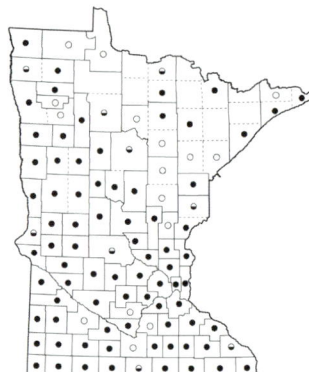

DISTRIBUTION

Boreal Chorus Frogs are found in the central United States and Canada. They range from the Northwest Territories and Ontario south to New Mexico, Kansas, Missouri, and Illinois (Moriarty Lemmon, Lemmon, and Cannatella 2007). In Minnesota, they are found statewide.

HABITAT

Boreal Chorus Frogs require temporary or permanent waters without fish populations. These wetlands are found in many moist habitats in Minnesota. Boreal Chorus Frogs venture into the uplands for a short distance, normally remaining within 100 meters (100 yards) of the breeding pond (Kramer 1974). This species is tolerant of urbanization and can be found in wetlands surrounded by development. Individuals apparently hibernate under rocks and logs adjacent to the water's edge (Vogt 1981).

Calling male Boreal Chorus Frog. Photograph by Allen Blake Sheldon.

LIFE HISTORY

Boreal Chorus Frogs are among the first frogs to emerge in the spring. They emerge from late March through early April. Breeding activity begins immediately and continues into early May. They are commonly found or heard calling from grassy wetlands. Their call is an ascending trill similar to the sound made when running one's thumb down the teeth of a comb. The speed of the one-to-two-second call varies with temperature, speeding up as the temperature becomes warmer. When approached, Boreal Chorus Frogs abruptly stop calling and will not resume for some time. The cessation of calling by one frog usually results in the entire chorus shutting down.

When they breed, Boreal Chorus Frogs lay small groups of eggs (5 to 20) attached to submerged vegetation. H. M. Smith (1934) found egg masses with up to 75 eggs. Individual females may lay hundreds of eggs over the

Boreal Chorus Frog egg mass. Photograph by Allen Blake Sheldon.

course of the breeding season (Pettus and Angleton 1967). The eggs hatch within 6 to 18 days (Vogt 1981), depending on the water temperature. Tadpoles transform in 8 to 10 weeks (Breckenridge 1944).

Boreal Chorus Frogs feed on small prey items, such as emerging aquatic insects, beetles, and other small insects. Whitaker (1971) found that they feed mainly on ants and spiders.

Boreal Chorus Frog, tadpole. Photograph by Allen Blake Sheldon.

REMARKS

In Oldfield and Moriarty (1994) two subspecies of Western Chorus Frogs were recognized as occurring in Minnesota, the Boreal Chorus Frog *(P. triseriata maculata)* and the Western Chorus Frog *(P. t. triseriata)*. Lemmon et al. (2007) reviewed the genus *Pseudacris* and reevaluated the species in Minnesota, elevating *P. maculata* to a full species. This analysis shifted the range of Western Chorus Frog *(P. triseriata)* outside of Minnesota.

Results of the Minnesota frog and toad call surveys indicate that Boreal Chorus Frog populations are stable (Larson 2009).

Boreal Chorus Frogs in vegetation. Photograph by Minnesota Department of Natural Resources—Carol D. Hall.

Family Ranidae—True Frogs

The family Ranidae, or true frogs, is another widespread family, with 371 species and representatives on all the continents except Antarctica. Generally true frogs have large eyes, tympanums (eardrums), and large, highly developed hind legs. The world's largest frog, the Goliath Frog of central Africa, is a member of this family. Minnesota's largest frog, the American Bullfrog, is also in the family Ranidae.

Within the Ranidae family, the species of North, Central, and South American frogs that were previously in the genus *Rana* were removed from this large and predominantly Eurasian taxon and assigned to the genus *Lithobates* (Frost et al. 2006; Hillis and Wilcox 2005).

Minnesota has six representatives of the family Ranidae: the American Bullfrog, the Green Frog, the Pickerel Frog, the Northern Leopard Frog, the Mink Frog, and the Wood Frog.

American Bullfrog

Lithobates catesbeianus

DESCRIPTION

The American Bullfrog is the largest frog in Minnesota. The green skin color of the American Bullfrog varies in shade depending on its size and the air temperature. Lighter shades of skin color are found in smaller frogs and during warm temperatures. Adults, especially males, are sometimes a mottled green and brown. The throat of adult males is yellow. Juveniles have small, dark spots and smooth skin, which becomes slightly bumpy or granular as they mature. The belly is white with a gray to yellow wash, especially under the chin, and is sometimes mottled with brown. Individuals can reach a snout–vent length of 20.3 centimeters (8 inches) (Conant and Collins 1998).

Unlike all other true frogs in Minnesota, American Bullfrogs lack a dorsolateral ridge. Males have large, convex tympanic membranes that are larger than their eyes, whereas females have flat tympanic membranes the same size as their eyes. Adult males also have an enlarged pad on their thumb.

American Bullfrog tadpoles are the largest tadpoles in the state. They are green with dark mottling. Black spots are scat-

American Bullfrog, adult female. Photograph by Allen Blake Sheldon.

tered over the upper body and tail. A mature tadpole can reach 12 centimeters (4¾ inches). The ratio of body length to tail length is 1:1.5.

DISTRIBUTION

The American Bullfrog is native to the region east of the Rocky Mountains, except the northern Great Plains. It has been introduced throughout the western United States and can now be found in nearly every state.

In Minnesota, the natural distribution of the American Bullfrog is limited to the extreme southeastern corner from Winona south along the Mississippi River backwaters. Introduced populations have been documented in Blue Earth, Chisago, Dodge, Douglas, Hennepin, Jackson, Mower, Nicollet, Rice, Stearns, Steele, Waseca, and Washington Counties.

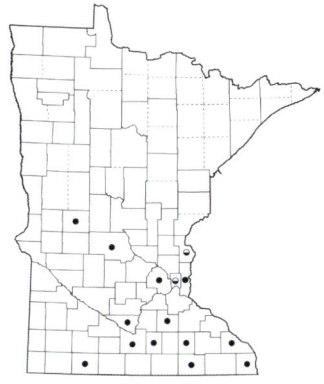

HABITAT

American Bullfrogs are rarely found far from the edge of permanent bodies of water, including rivers, lakes, ponds, and ditches. They use areas with thick emergent vegetation and require open access to water. American Bullfrogs are commonly found along the sloughs and backwaters of the Mississippi River in Houston County.

They are able to coexist with game fish and breed in a variety of permanent aquatic habitats. Adults and tadpoles overwinter underwater in lakes and rivers that are sufficiently deep to avoid oxygen depletion and freezing.

Head of adult male American Bullfrog. Photograph by Allen Blake Sheldon.

LIFE HISTORY

American Bullfrogs breed later than other frogs in Minnesota. They emerge from winter dormancy in early May and feed before their breeding season. Males start calling the well-known deep baritone "jug-a-rum" in June and continue into mid-July. They establish a territory along the shore of a lake or

river, defending it from other males. When they encounter calls of unfamiliar males, they defend their territory more aggressively (Bee and Gerhardt 2002). Females lay up to 20,000 eggs (Johnson 2000) in a large surface film. The tadpoles take two to three years of development before undergoing metamorphosis (Wright 1920), the longest time of any Minnesota frog. The active season in Minnesota is not long enough for full development of the tadpoles in one or two summers. The young frogs then take from two to five years to become sexually mature (DeGraaf and Rudis 1983). Growth and maturation are slower at colder temperatures.

American Bullfrogs have small home ranges, the radius varying from 0.6 to 11.3 meters (2 to 37 feet) and averaging 2.6 meters (8 feet) (Bury and Whelan 1984). Their main defense is to leap away from shore in a series of quick jumps over the water surface (Vogt 1981). They often produce a short squawk as they jump into the water. If caught, they may give a distress call that is suggestive of a cat scream.

American Bullfrogs eat anything that can fit into their

American Bullfrog egg mass. Photograph by Allen Blake Sheldon.

American Bullfrog, tadpole. Photograph by Allen Blake Sheldon.

mouths. Their diet is primarily insects but includes earthworms, frogs, small turtles, snakes, bats, mice, and ducklings (Bury and Whelan 1984). American Bullfrogs in turn are preyed on by humans, raccoons, mink, and large fish.

REMARKS

Bullfrogs are a game species in Minnesota and are hunted for their legs, which are considered a delicacy. Collection is regulated by the DNR's Section of Fisheries through the use of seasons and bag limits.

In Oldfield and Moriarty (1994), the American Bullfrog was listed as Bullfrog *(Rana catesbeiana)*. *American* was added to the name to separate it from other species of bullfrogs (Crother 2012). The large genus *Rana* was also split, and all of Minnesota's species were placed in the genus *Lithobates* (Frost et al. 2006). The species name was changed to *catesbeianus* to reflect the gender shift in the genus name.

Outside of Houston and Winona Counties American Bullfrogs are considered invasive. If left unchecked, the American Bullfrog could eventually displace other native species in aquatic habitats throughout Minnesota.

In Minnesota, several new records of American Bullfrogs have been documented in counties far outside their native range. These are likely the result of the release of frogs or tadpoles purchased from the pet trade or backyard pond suppliers. In addition, teachers have occasionally purchased tadpoles for classrooms and released the transformed frogs. Neither frogs nor tadpoles should be released in Minnesota.

Green Frog

Lithobates clamitans

DESCRIPTION

Green Frogs are medium-sized frogs with skin color ranging from green to brown. Young frogs generally have a blotched skin pattern. The upper lip of the Green Frog is a lighter and brighter green than the head. The belly is white, and adult males usually have a yellow throat. Dorsolateral ridges are typically present, although they may be reduced and less obvious in large adult males. In adult females the size of the tympanum is equal to the diameter of the eye; in adult males the tympanum is larger than the diameter of the eye.

Green Frogs range in snout–vent length from 6 to 9 centimeters (2⅜ to 3½ inches). Some individuals, however, can reach over 10.8 centimeters (4¼ inches) (Conant and Collins 1998). Special attention is needed for correct identification where

Green Frog, adult female. Photograph by Allen Blake Sheldon.

Green Frogs and Mink Frogs are found together in Minnesota. Mink Frogs give off a rotten odor when their skin is scratched, whereas Green Frogs do not.

Tadpoles of Green Frogs are 7.4 to 10 centimeters (3 to 4 inches) in total length and have a ratio of body length to tail length of 1:1.8. A few dark marks are present on the body, in contrast to the tail and tail fin, which are strongly mottled.

DISTRIBUTION

Green Frogs are found east of the Great Plains from the Gulf of Mexico north to central Ontario and Quebec. In Minnesota, they are found in the eastern half of the state.

HABITAT

Green Frogs occur in wetlands with permanent water and emergent vegetation, including rivers, streams, and their associated backwaters (Fleming 1976). Lakes and ponds with shallow margins and springs and seeps with permanent water are also used. Green Frogs require aquatic sites that do not freeze solid and that maintain enough oxygen for overwintering tadpoles and adults.

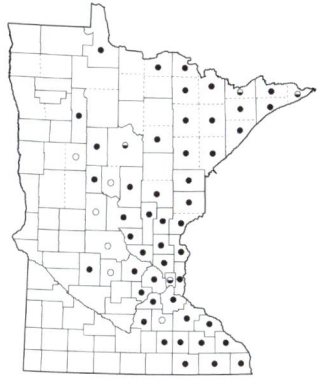

Albino juvenile Green Frog. Photograph by John J. Moriarty.

Green Frog egg mass. Photograph by Allen Blake Sheldon.

LIFE HISTORY

Green Frogs emerge from winter dormancy in late April (Walker 1946). They breed from late May to mid-August in the same area as they overwinter (Breckenridge 1944). The males call with a single "plunk," a sound similar to a plucked, out-of-tune banjo string. They defend a territory that ranges from 12 to 21 square meters (14⅓ to 25 square yards) (Hamilton 1948). Females lay up to 4,000 eggs per year in large floating masses. Most tadpoles overwinter and transform throughout the following summer. Newly metamorphosed frogs are terrestrial and spend their first summer on land (Vogt 1981), thus reducing competition for food resources with adults. Young frogs mature in their second year after metamorphosis (Wells 1977).

Adult Green Frogs have an average home range of 61 square meters (73 square yards), but home ranges can be as large as 200 square meters (240 square yards) (Hamilton 1948). These frogs are highly aquatic and rarely venture from water after their first summer. In northern Minnesota, they regularly occur in association with Mink Frogs. Green Frogs are found along the water's edge, while Mink Frogs prefer floating vegetation in deeper water. This partitioning of habitat keeps the two species from competing for the same food resources (Fleming 1976).

When disturbed, Green Frogs dive from shore into the water and swim to the bottom to elude predators. The frogs often produce a short alarm call before escaping into the water. Large fish (e.g., bass and northern pike) and Common Watersnakes regularly feed on them. Green Frogs prey on insects and occasionally on frogs, crayfish, and small fish.

REMARKS

Minnesota has a population of Green Frogs in Pine County that produce a high percentage of albino tadpoles. No albino adults have been reported. Several tadpoles were raised in captivity until metamorphosis, but the transformed frogs did not survive (J. Moriarty, pers. obser.). Normal tadpoles from the same pond transformed and survived.

Shoreline development had negative impacts on Green Frog populations in Wisconsin (Woodford and Meyer 2003), where fewer adult frogs were present along shorelines with less vegetative cover.

In Oldfield and Moriarty (1994), Green Frogs were listed as *Rana clamitans*. In 2006, the genus *Rana* split, and all of Minnesota's species were placed in the genus *Lithobates* (Frost et al. 2006).

Green Frog, tadpole. Photograph by Allen Blake Sheldon.

Pickerel Frog

Lithobates palustris

DESCRIPTION

Pickerel Frogs are medium-sized tan frogs with two distinctive parallel rows of dark-brown, squarish spots on their back and a bright-yellow or orange coloring in the groin area. Males have vocal sacs on each side of the head, which are evident when they call, and they develop dark, swollen thumbs during the breeding season. Snout–vent length of Pickerel Frogs ranges from 4.5 to 8 centimeters (1¾ to 3¼ inches). They are smaller than similar-looking Northern Leopard Frogs. Northern Leopard Frogs may also have a light-yellow wash in the groin area, but the spots on their backs are roundish and randomly distributed.

Pickerel Frog tadpoles have a low dorsal fin and a wide ventral fin. The body has scattered mottling, which becomes heavier on the tail. Their total length ranges from 4.8 to 5.5 centimeters (1⅞ to 2⅛ inches), and the ratio of body length to tail length is 1:1.7.

Pickerel Frog, adult. Photograph by Allen Blake Sheldon.

DISTRIBUTION

Pickerel Frogs are found throughout the eastern United States and southern Great Plains with the exception of most of Illinois and the extreme southeastern United States. They are restricted to the southeast corner of Minnesota.

HABITAT

Pickerel Frogs are found along the edges of small ponds, mid-sized rivers, and spring-fed streams, in both wooded and open areas. During the summer, they venture into wet meadows, especially those adjacent to streams, though they rarely venture far into uplands, as Northern Leopard Frogs do; Pickerel Frogs have been found on bluff prairies in southeastern Minnesota at least 500 meters (547 yards) from water. Pickerel Frogs overwinter in streams and ponds. In the southern United States they overwinter in caves, sometimes even in crevices in the ceiling.

LIFE HISTORY

Emergence occurs in April, followed by the breeding season, which continues through May (Vogt 1981). Males call with a soft, low-pitched snore during the day as well as at night. Because they regularly call while totally submerged, their voice has limited carrying capacity (Barbour 1971). Jeff LeClere (pers. comm.) has recorded clicking sounds, similar to static, produced between the typical snore-like breeding call. Females lay 200 to 3,000 eggs and

Pickerel Frog, tadpole. Photograph by Allen Blake Sheldon.

Pickerel Frog habitat in southeast Minnesota. Photograph by Allen Blake Sheldon.

attach them to vegetation and twigs (Wright and Wright 1949). The eggs usually hatch in two weeks (DeGraaf and Rudis 1983). Tadpoles metamorphose in 80 to 100 days but may overwinter if the breeding occurs late in the season.

The diet of Pickerel Frogs is made up mainly of terrestrial arthropods and insects, including beetles and grasshoppers that venture too close to the water's edge.

Pickerel Frogs generally stay close to water year-round. Their proximity to water is advantageous in escaping predators. Another defense mechanism is toxic skin secretions. Pickerel Frogs are very distasteful to predators, and their toxic secretions can kill other amphibians, including other Pickerel Frogs, that are placed together in the same container.

REMARKS

The Pickerel Frog is considered a Species of Greatest Conservation Need by the Minnesota DNR (2006).

In Oldfield and Moriarty (1994), Pickerel Frogs were in the genus *Rana*. The genus *Rana* split, and all of Minnesota's species were placed in the genus *Lithobates* (Frost et al. 2006).

Northern Leopard Frog
Lithobates pipiens

DESCRIPTION

The Northern Leopard Frog is familiar to most Minnesotans. Skin color is green or brown with two or three rows of irregularly sized, dark spots on the back. Dorsolateral ridges are prominent. The legs have dark bars, and the belly is white. Occasionally there is a light-yellow wash on the underside of the legs. Juvenile coloration is similar to adult. The snout–vent length of this medium-sized frog ranges from 5 to 9 centimeters (2 to 3½ inches); the record length is 11.1 centimeters (4⅜ inches) (Conant and Collins 1998). Males have paired vocal pouches and develop swollen thumbs with dark-purple pads during the breeding season.

In addition to the normal color pattern, two variants of the Northern Leopard Frog are found in Minnesota, the Burnsi and Kandiyohi morphs. The Burnsi morph lacks spots, although the legs may be barred in some individuals. The Kandiyohi morph has extra black pigment between spots. These two morphs were

Northern Leopard Frog, adult. Photograph by Allen Blake Sheldon.

originally described as separate species (Weed 1922), but later work revealed that they were both *Lithobates pipiens,* differing from normal Northern Leopard Frogs by a single dominant gene (Breckenridge 1944). The Kandiyohi color morph is very rare; only two or three individuals are found per year (Hoppe and McKinnell 1989, 1991a). The Burnsi color morphs are more abundant and can be found in bait shops from time to time.

Tadpoles of Northern Leopard Frogs have a dark back with a cream belly and a tail that is lighter in color than the body. They range in size from 4.5 to 8.4 centimeters (1¾ to 3⁵⁄₁₆ inches) with a ratio of body length to tail length of 1:1.5. They are similar to Green Frog tadpoles but darker.

DISTRIBUTION

In the United States, Northern Leopard Frogs are found from the Great Basin east, through the north-central United States, into New England. In Canada they extend from Nova Scotia to the Northwest Territories. Northern Leopard Frogs are found statewide in Minnesota. They are the only amphibian with records from every county. The Burnsi morph occurs more commonly in the east-central part of the state, while the Kandiyohi morph occurs mostly in west-central Minnesota (Hoppe and McKinnell 1991a, 1991b).

Hibernating Northern Leopard Frog. Photograph by Minnesota Department of Natural Resources—Bernard Sietman.

Calling male Northern Leopard Frog. Photograph by Allen Blake Sheldon.

Northern Leopard Frog egg mass. Photograph by Allen Blake Sheldon.

HABITAT

Northern Leopard Frogs use a wide range of habitats, but they are most abundant in wet meadows and open fields adjacent to ponds and lakes. They prefer meadows with grasses 15 to 30 centimeters (6 to 12 inches) tall. The grasses are sufficiently tall to provide cover for hiding but not so tall as to interfere with their movements. Insects, a valuable food source, are common and accessible in these areas (Merrell 1977). Northern Leopard Frogs may travel up to 1.5 kilometers (0.9 mile) from water if there is adequate moisture and humidity in the grassy cover. Overwintering occurs in water, where they may be found in the aquatic vegetation of ponds and streams. Large congregations have been found at the inflow to ponds, where currents provide adequate oxygen.

LIFE HISTORY

Individuals become active as water temperatures begin to rise in the spring. By the time of the spring thaw, Northern Leopard Frogs are moving overland to breeding ponds. In the spring, overland movements occur more often during the day, since evening temperatures are too cool (Merrell 1977). On rainy nights

Northern Leopard Frog, tadpole. Photograph by Allen Blake Sheldon.

in early fall, large migrations of Northern Leopard Frogs can be seen crossing Minnesota roads. They may travel approximately 100 meters (109 yards) each night from mid-September to the end of October during prehibernation migrations to ponds and streams (Dole 1965).

Male Northern Leopard Frogs begin calling when water temperatures are above 20°C (68°F) and after they reach the breeding ponds in April. The call is a low snore mixed with resonant grunts and squeaks. Their advertisement call may be the most complex of any North American temperate-zone anuran (Larson 2004). At the peak of the breeding season, the males attempt amplexus with other males or anything else that is in the way, including floating beer cans (Merrell 1977). Females are never abundant in the breeding ponds because they stay only

Northern Leopard Frog in Burnsi phase. Photograph by Allen Blake Sheldon.

long enough to lay their eggs (Merrell 1970). Females may lay up to 6,500 eggs; in Minnesota, Merrell (1965) has reported a range of 2,000 to over 5,000 eggs. Eggs are laid in globular masses, often concentrated in one area of the breeding pond. Tadpoles metamorphose in approximately three months after the eggs are laid but may not metamorphose the first year if breeding is late or the summer is cool. Juveniles leave the water and head to grassy areas to feed on terrestrial insects. Juveniles spend most of their time closer to water than do adults.

Northern Leopard Frog, metamorph. Photograph by Allen Blake Sheldon.

Northern Leopard Frog in Kandiyohi phase. Photograph by James E. Gerholdt.

Adults feed primarily on insects and occasionally on frogs, invertebrates, or snails. They take only prey that is moving (Merrell 1977). Predators that feed on Northern Leopard Frogs include herons, raccoons, snakes, and owls. Humans use large numbers of these frogs for fishing bait. Additionally, cars and mowers kill untold numbers.

REMARKS

In Oldfield and Moriarty (1994), Northern Leopard Frogs were in the genus *Rana*. The genus *Rana* split, and all of Minnesota's species were placed in the genus *Lithobates* (Frost et al. 2006).

Until recently the leopard frog was thought to be a single, wide-ranging species found throughout the United States except the West Coast (Conant 1958). In the 1970s the leopard frog complex was split into four species (Pace 1974). Today, the Southern Leopard Frog *(Lithobates utricularia)*, Plains Leopard Frog *(Lithobates blairi)*, and Rio Grande Leopard Frog *(Lithobates berlandieri)* are recognized as different species and have a combined range from Texas to Nebraska and east to Florida and New Jersey.

Northern Leopard Frogs are regulated by the DNR's Section of Fisheries. A fishing license is required for collecting. Approximately 1,450 kilograms (3,200 pounds; over 50,000 frogs) were sold for nonbait purposes, mainly to biological supply houses, in 1989 (DNR Fisheries, unpublished data). The capture of frogs for bait takes an even larger number of frogs per year, but the DNR does not keep records.

Annual call surveys in Wisconsin indicate a continuous decline in Northern Leopard Frog populations since 1984 (Kitchell and Bergeson 2010). No trends were detected in Northern Leopard Frog populations based on Minnesota call surveys (Larson 2009).

Mink Frog

Lithobates septentrionalis

DESCRIPTION

Mink Frogs are frequently seen sitting on lily pads in lakes and ponds in northern Minnesota. Skin color is green with brown mottling on the dorsal surface, and the belly color ranges from white to light yellow with gray mottling. The sides of the head and lips are bright green. Dorsolateral ridges vary in prominence among individuals, ranging from very prominent to faint. A distinguishing characteristic of the Mink Frog is a skin odor similar to the smell of rotten onions, which is given off when the frog's back is scratched. The size of a male's tympanum is larger than its eye, whereas that of a female's is equal to or smaller than its eye. The snout–vent length of this small- to medium-sized frog ranges from 5 to 7 centimeters (2 to 2¾ inches). Females tend to be larger than males (Hedeen 1970).

Mink Frogs appear similar to Green Frogs but can be distinguished readily on close examination. If the rotten onion odor

Mink Frog, adult. Photograph by Allen Blake Sheldon.

is not evident, then the hind foot should be examined. The webbing on the hind feet of Mink Frogs extends all the way to the tip of the fifth toe; the webbing only goes to the second digit on Green Frogs.

Mink Frog tadpoles are brilliant green or yellowish green with black spots and a reddish tail fin (Vogt 1981). They are 2.5 to 3.5 centimeters (1 to 1½ inches) in length and are similar in size and shape to Green Frog tadpoles.

DISTRIBUTION

The Mink Frog is found along the northern edge of the United States from Minnesota to northern New England and adjacent Canada. The species is found in the northern half of Minnesota including the northeast.

HABITAT

Ponds, lakes, and slow-moving rivers throughout the forested areas of the state are the primary habitats utilized by Mink Frogs. Individuals are often seen sitting on water lilies or other emergent vegetation, including floating sphagnum borders of bogs. Mink Frogs spend the winter underwater in ponds, streams, and bogs.

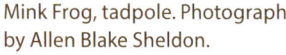

Mink Frog, tadpole. Photograph by Allen Blake Sheldon.

Mink Frog in habitat. Photograph by Minnesota Department of Natural Resources—Carol D. Hall.

LIFE HISTORY

Mink Frogs emerge from winter dormancy in May (Harding 1997). Males begin their calling in late May, and the breeding season extends from late June into early August. Males float in the water while making a "knock-knock-knock" vocalization. Although males interact aggressively when encountering other males, Bevier et al. (2006) found that they did not defend territories over prolonged periods, as do Green Frogs and American Bullfrogs. The eggs are laid in a globular mass that contains 500 to 4,000 eggs. Vogt (1981) reports that eggs are readily laid in water 1 meter (3 feet) deep or deeper. The length of time needed prior to hatching appears to be variable; this aspect of the Mink Frog's life history needs more research. Tadpoles metamorphose in one year, but some may take two years (Hedeen 1970).

The Mink Frog is the most aquatic frog in Minnesota (Schmid 1965). Hedeen (1970) found that during the summer individuals rest with only their eyes and nose out of the water. In contrast, Green Frogs remain near the shore (Fleming 1976). Mink Frogs move out on land on humid nights or during rain events. When they are sitting on water lilies, they can easily be approached from canoes.

Insects and their larvae comprise the bulk of the Mink Frog's diet, although the frogs do take minnows and other aquatic invertebrates (Hedeen 1970). Predators that feed on other frogs, such as raccoons, striped skunks, and gartersnakes, also feed on Mink Frogs. Their musky odor may discourage some predators.

REMARKS

In Oldfield and Moriarty (1994), Mink Frogs were in the genus *Rana*. The genus *Rana* split, and all of Minnesota's species were placed in the genus *Lithobates* (Frost et al. 2006).

Minnesota may support the largest population of Mink Frogs in the United States (Anderson and Baker 2002). Although populations appear stable in Minnesota, annual call surveys in Wisconsin show a steady decline between 1984 and 2010 (Kitchell and Bergeson 2010).

Wood Frog

Lithobates sylvaticus

DESCRIPTION

The Wood Frog sports a distinctive black face mask. Skin color ranges from light tan to dark brown, and a pinkish tinge is sometimes present. Solid or dashed dark-brown lines may follow along the prominent dorsolateral folds. Some individuals from northern and western Minnesota have a narrow, white middorsal stripe. The belly is white without markings. They are small- to medium-sized frogs with a snout–vent length that ranges from 5 to 7 centimeters (2 to 2¾ inches); adult females are slightly larger than adult males. The male's vocal sacs are paired and evident during the breeding season, even when deflated (Fishbeck 1968).

Wood Frog tadpoles have a greenish-brown body without distinct markings, although the rounded tail fins have faint mottling. The tadpoles are 4.2 to 4.8 centimeters (1⅝ to 1⅞ inches) long, and the ratio of body length to tail length is 1:1.8.

Wood Frog, adult. Photograph by James E. Gerholdt.

DISTRIBUTION

Wood Frogs are found in the northeast and north-central United States, ranging south to Georgia and Tennessee in the Appalachian Mountains. They are found across most of Canada west to Alaska and north of the Arctic Circle. They are found farther north than any other frog species in the Western Hemisphere. In Minnesota, Wood Frogs are found statewide except for the southwestern corner. Their distribution is spotty in the southeastern corner.

HABITAT

Wood Frogs are found in moist forests, both deciduous and coniferous. In prairie areas they may be found in bottomland forests along rivers. Marshes or small ponds (permanent or temporary) are required for breeding sites. In Itasca State Park, Fishbeck (1968) found juvenile Wood Frogs living in sphagnum moss, which provides good cover from predators. Winter is spent partially frozen under leaf litter on the woodland floor (Schmid 1982).

Wood Frogs in amplexus. Photograph by Allen Blake Sheldon.

Calling male Wood Frog.
Photograph by Allen Blake
Sheldon.

LIFE HISTORY

Wood Frogs are one of the earliest frogs to breed in Minnesota, starting as soon as the ice disappears from small ponds in late March or early April. In some years breeding is postponed up to a week because of a cold spell or snow (Fishbeck 1968). Flooded alder stands or meadows with standing water are regularly used for egg laying during the short breeding season, which begins and ends in less than two weeks.

At breeding ponds males call while floating on the water's surface, vocalizing during the day or night. Their loud choruses act like beacons that attract other males to these short-lived breeding aggregations (Bee 2007). Males grab anything that moves. They can form "mating balls" consisting of a number of males grappling for a single female, sometimes drowning the female. The call is a duck-like quacking, and a chorus of frogs sounds like a group of feeding mallards. Large globular masses of eggs (500 to 800) may be attached to vegetation or left floating on the surface. Fifty to 100 egg masses are usually communally

Wood Frog egg mass.
Photograph by Allen Blake
Sheldon.

deposited in a small area at one end of the pond. Karns (1984) found deposition sites with over 400 egg masses in Koochiching County peatlands. Waldman (1982) found that these large, black, floating egg mats have higher temperatures than the surrounding water, thus providing a warmer environment for egg development. Eggs hatch in 14 to 20 days depending on the water temperature. Tadpoles metamorphose after six to nine weeks, and the young frogs move into the woods. Summer activity is dependent on weather. During dry periods, Wood Frogs spend long inactive periods in leaf litter or under logs.

Wood Frog, tadpole. Photograph by Allen Blake Sheldon.

Wood Frog breeding habitat. Photograph by Minnesota Department of Natural Resources—Carol D. Hall.

The diet of Wood Frogs is varied but primarily consists of flies and beetles (Vogt 1981). Wood Frogs are preyed on by gartersnakes, birds, and raccoons. Vogt (1981) found mink feeding on them at breeding ponds.

REMARKS

In Oldfield and Moriarty (1994), Wood Frogs were in the genus *Rana*. The large genus *Rana* was split, and all of Minnesota's species were placed in the genus *Lithobates* (Frost et al. 2006). The species name was changed to *sylvaticus* to reflect the gender shift in the genus name. Wood Frogs are the most common amphibians in the peatlands of northern Minnesota, and this species demonstrates a high tolerance for acidic waters (Karns 1992a).

Wood Frogs breed successfully in peatland waters with a pH as low as 4.5 (Karns 1992b). This ability to withstand acidic conditions may allow Wood Frogs to survive in areas with increased acid precipitation and runoff.

An active Wood Frog was observed in January 2012 moving near a pond in a forested metropolitan park after an unseasonal rain event. In a typical year males that arrive early to breeding ponds may have an advantage; however, such unprecedented fluctuations in weather patterns may result in high mortality if large numbers of individuals have initiated migration.

Family Ambystomatidae— Mole Salamanders

The Mole salamander family is comprised of roughly 30 species, all restricted to North America. Of these, 16 species are found within the United States, including 4 in Minnesota. These salamanders are stout bodied with short, wide heads and four well-developed limbs. There are four toes on the front feet, and five on the hind feet. Generally the costal grooves are clearly visible. Eggs are laid in the water. Gill-bearing larvae typically transform into terrestrial adults that spend most of their lives under logs or underground in mammal tunnels. Given their fossorial tendency, terrestrial adults are most frequently observed during spring breeding activities and fall migrations. To survive variability in landscape and weather patterns, some species within this group have developed adaptations, including neoteny (becoming sexually mature in the larval form) and parthenogenesis (reproduction occurring without a fertilized egg).

The Blue-spotted Salamander, Spotted Salamander, Western Tiger Salamander, and Eastern Tiger Salamander are the representatives of this family in Minnesota.

Blue-spotted Salamander

Ambystoma laterale

DESCRIPTION

The Blue-spotted Salamander is a small woodland salamander that is relatively common in Minnesota's forested habitats. Adults have a black, blue-black, or grayish-black dorsal surface with a somewhat lighter belly. The sides are speckled with small bluish-turquoise spots, similar to vintage enamel cookware. Some spots extend onto the back and belly. Occasionally only a few spots are present or are totally absent. These salamanders have 12 costal grooves and attain a length of 7 to 13 centimeters (3 to 5 inches) (Conant and Collins 1998). Males tend to be smaller than females, and during the breeding season males have swollen cloacas. Newly transformed larvae have yellow speckles instead of blue but otherwise resemble adults.

Blue-spotted Salamander, adult. Photograph by Jeffrey B. LeClere.

DISTRIBUTION

Blue-spotted Salamanders were invaders of postglacial habitats (Holman 1998) and represent the northernmost salamander in North America. Today populations are distributed around the Great Lakes from the extreme northeastern United States and southern Canada, west to Iowa, Minnesota, and Ontario. In Minnesota, this species commonly occurs in northeast and east-central portions of the state. Isolated populations also occur in riparian corridors and forest remnants in southeastern Minnesota.

HABITAT

These salamanders inhabit the forest floor of moist woodlands and boreal forests. They are often associated with sandy soil (Vogt 1981). Small ponds and woodland pools that retain water until late summer are used for breeding (Wilbur and Collins 1973). Backwaters of slow-moving streams also provide breeding habitat. Blue-spotted Salamanders are absent where pH is less than 4.5 (Karns 1992a, 1992b).

During the summer adults may be found taking shelter under moist, decaying logs, bark, or moss. They are not tolerant of freezing conditions and need to move below the frost line as winter progresses (Storey and Storey 1986).

Highly spotted Blue-spotted Salamander. Photograph by Allen Blake Sheldon.

Blue-spotted Salamander eggs. Photograph by Allen Blake Sheldon.

LIFE HISTORY

This species breeds shortly after the ice has melted from shallow forest wetlands. They migrate to wetlands at roughly the same time that Wood Frogs and Boreal Chorus Frogs breed. Courtship activity has been observed as early as April 15 in Wisconsin (Vogt 1981) and can last from a few days to a few weeks depending on weather conditions (Petranka 1998). Most breeding activity occurs at night, and the elaborate courtship displays involve the male clasping the female from above while rubbing his nose over the front of her body. Periods of thrashing about in the water are followed by the pair lying quietly on the pond bottom. After these courtship encounters cease, the male climbs off the female and deposits from 8 to 35 spermatophores on sticks, leaves, or other structures in the water (Petranka

Blue-spotted Salamander, larva. Photograph by Allen Blake Sheldon.

Blue-spotted Salamander habitat on Lake Superior. Photograph by Minnesota Department of Natural Resources—Carol D. Hall.

1998). The female picks these up with her cloacal lips, and fertilization occurs internally. Eggs are laid singly or in small clusters of 6 to 10 at the bottom of the pond and are attached to vegetation. Hatching occurs in three to four weeks, and metamorphosis occurs after an additional two to three months (Talentino and Landre 1991). Larvae feed primarily on small aquatic invertebrates (Nyman 1991) but may also eat small tadpoles (C. K. Smith 1983). Metamorphosis typically occurs in August (Pfingsten and Downs 1989); however, development is dependent on climate and is delayed in colder conditions. Larvae have been observed in late September in rock pools adjacent to Lake Superior in northeastern Minnesota.

Adult Blue-spotted Salamanders feed on a variety of invertebrates, including beetles, snails, earthworms, sow bugs, and spiders (Minton 1972).

REMARKS

Blue-spotted Salamanders utilize various forms of defense, including arching their tail and waving it slowly back and forth when approached by a predator (Brodie 1977). An apparently foul-tasting, sticky substance is excreted from glands at the base of the waving tail (Pfingsten and Downs 1989).

Unisexual hybrids of Blue-spotted and Jefferson Salamanders (*Ambystoma jeffersonianum*) occur in Minnesota, although no additional specimens have been confirmed outside of the original Cass County record (Dorff 1995c). Additional hybrid populations occur in the Great Lakes region, including northern Wisconsin (Vogt 1981); however, the closest Jefferson Salamander population is in eastern Illinois. Since hybrids closely resemble Blue-spotted Salamanders, they are easily overlooked and may be more widely distributed in Minnesota's forested region. In comparison to Blue-spotted Salamanders, the unisexuals tend to be larger with a lighter base color and small flecking instead of distinct spots. Female hybrids mate with male Blue-spotted Salamanders to stimulate development of the egg; however, the male's nuclear DNA makes no genetic contribution to the offspring. Although formerly referred to as Tremblay's Salamander *(Ambystoma tremblayi),* this triploid form is not considered a unique species (Phillips and Mui 2005).

Threats include forest fragmentation and acidification of breeding ponds (< 4.5 pH). Limited dispersal abilities put small, isolated populations at risk. In areas dominated by agricultural fields, such as south-central and southeastern Minnesota, forested corridors along rivers and streams may provide some of the only remaining habitat for this species. Woodland ponds within these corridors provide critical breeding habitat for these isolated populations.

Forested zones that surround breeding sites have been described by Semlitsch (1998), who recommends retaining core habitat with a radius of 164 meters (538 feet) around breeding sites to protect all life stages of pond-breeding salamanders.

Spotted Salamander
Ambystoma maculatum

DESCRIPTION

Spotted Salamanders are medium- to large-sized salamanders with a dark-colored background and two rows of yellow spots running down their back and tail. Their lower sides and belly are gray, often with white flecks along their sides. Their head is rounded with spots on the back of the head ranging in color from yellow to orange. They have from 11 to 13 costal grooves, and adults range from 15 to 25 centimeters (5.9 to 9.8 inches) in total length (Petranka 1998). Females are generally larger than males. During the breeding season males have a swollen vent.

DISTRIBUTION

In Minnesota, the Spotted Salamander was initially documented in 2001 and is currently known to occur in eastern Pine and Carlton Counties (Hall 2002). This species can be found in southern Canada, south through most of the eastern United States and as far west as Minnesota and eastern Texas.

Spotted Salamander, adult. Photograph by Allen Blake Sheldon.

HABITAT

In Minnesota Spotted Salamanders occupy mature forest habitats, often dominated by deciduous species. Seasonal, fishless wetlands provide habitat for breeding.

LIFE HISTORY

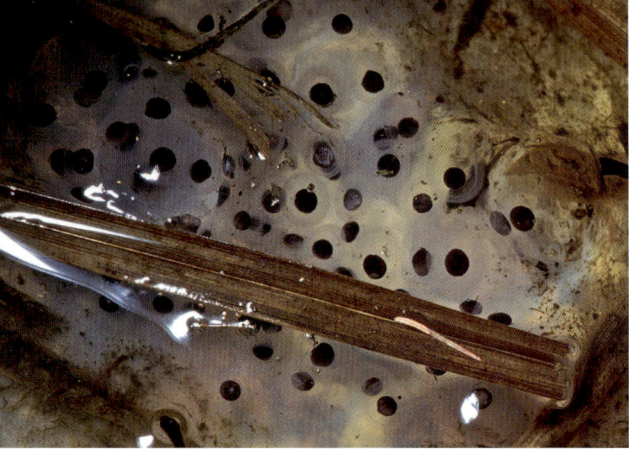

Spotted Salamander eggs. Photograph by Allen Blake Sheldon.

Spotted Salamanders move to breeding ponds in late April or early May, typically to fishless, seasonal ponds within upland forest habitat. Movements are triggered by a combination of heavy spring rain and an increase in the soil and air temperatures (Sexton, Phillips, and Bramble 1990). Males typically migrate to breeding sites before females (Sexton et al. 1986). Nonbreeding individuals may also migrate to breeding ponds (Shoop 1967). The male-biased sex ratio typical at breeding ponds can result in extreme competition among male Spotted Salamanders during their short breeding season (Flageole and Leclair 1992). Courtship involves nosing and writhing of bodies prior to the male depositing spermatophores on leaves and twigs in the pond (Vogt 1981). Males may "cap" spermatophores two to six high (Arnold 1976), presumably giving the top male the advantage when females pick up the top sperm mass. Fertilization takes place internally after the female picks up a sperm packet with her cloaca. Two to three days after breeding, females attach firm, gelatinous egg masses to sticks or aquatic vegetation, often near the periphery of the wetland. Egg

Spotted Salamander, juvenile. Photograph by Allen Blake Sheldon.

Spotted Salamander, larva. Photograph by Allen Blake Sheldon.

masses typically contain 75 to 100 eggs. The embryos develop within four to seven weeks, hatching into larvae that transform within two to four months. Transformed salamanders remain near the outer perimeter of the pond for two to three weeks before dispersing into the surrounding forest in late August or September. In some populations, slow-growing larvae may overwinter in the wetland, transforming the following spring (Bleakney 1952; Whitford and Vinegar 1966).

After breeding, Spotted Salamanders exit the wetland near

Spotted Salamander breeding habitat. Photograph by Minnesota Department of Natural Resources—Carol D. Hall.

the site where they entered it, suggesting that they follow the same route returning to their home territory (Shoop 1965, 1968; Phillips and Sexton 1989). Indeed, some salamanders were found returning to the same home burrow (Downs 1989). Possible migratory cues include olfaction (McGregor and Teska 1989; Whitford and Vinegar 1966).

REMARKS

Spotted Salamander egg masses are unique due to their large, gelatinous character but could be confused with fist-sized egg masses of the Wood Frog or Eastern Tiger Salamander. Spotted Salamanders have 75 to 100 eggs per mass compared to the Wood Frog's 500 or more eggs per mass. Also, Wood Frogs often lay their eggs communally.

The rows of spots on a Spotted Salamander's back were found to be an indicator of salamander genetic fitness. Salamanders from less fragmented habitat had rows with similar numbers of spots per row, indicating that there was less stress associated with these individuals (Davis and Maerz 2007).

The Spotted Salamander is listed as a species of special concern in Minnesota and is considered a Species of Greatest Conservation Need (Minnesota DNR 2006, 2013). Their primary threats in Minnesota include forest and wetland habitat loss. Additional threats include wetland acidification and high levels of aluminum, copper, zinc, and silica in breeding ponds (Savage and Zamudio 2005).

Western Tiger Salamander

Ambystoma mavortium

DESCRIPTION

Formerly considered a subspecies of the Eastern Tiger Salamander, this large salamander is more common in the Great Plains states, reaching the edge of its range in western Minnesota. It has a broad head, small eyes, and a long tail. Its markings are highly variable in both color and pattern, and the adult form can be terrestrial or aquatic. The black, brown, or olive background contrasts with markings that vary from bright yellow to dull brown or green. These markings occur across the dorsal surface, head, tail, and limbs and vary in size from small dots to large, bold patterns. In some forms the markings are small, dark spots on an olive background. The ventral surface is similarly marked but can be lighter in color with less evident spots on some adults and juveniles. Costal grooves range from 11 to 14. Four toes are present on each front foot, and five on each hind foot. Small hard tubercles are present on the soles of their feet to aid in digging. In Minnesota, adult Western Tiger Salamanders have been recorded as large as 350 millimeters (13.7 inches). During the

Western Tiger Salamander, adult. Photograph by Allen Blake Sheldon.

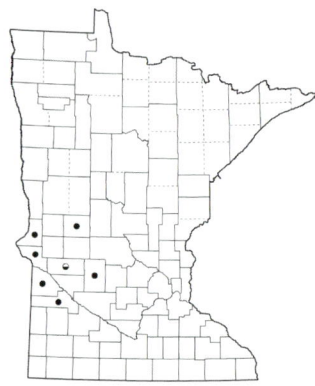

breeding season mature males have a swollen vent with papillae present at the rear of the vent.

Western Tiger Salamanders have a greater paedomorphic tendency (ability to retain gills as adults) than other species within the *Ambystoma* complex. In 2007 DNR Fisheries staff collected sexually mature, gilled Tiger Salamanders from wall-eye rearing ponds in several west-central Minnesota counties. Given the mature form of these individuals, they have been tentatively mapped as *A. mavortium*. However, DNA analysis currently in process will confirm the identity of these individuals.

DISTRIBUTION

The Western Tiger Salamander reaches the eastern edge of its range in Minnesota. This member of the *Ambystoma* complex occurs in central Alberta and Saskatchewan, south through the Great Plains to New Mexico and Texas (Petranka 1998).

Due to recent taxonomic changes in the *Ambystoma* complex and similarities in their morphological features, the distribution map of the Western Tiger Salamander in Minnesota is considered incomplete. Confirmed specimens collected from

A paedomorphic form of the Western Tiger Salamander. Photograph by Minnesota Department of Natural Resources—Carol D. Hall.

Western Tiger Salamander larvae showing variation. Photograph by Minnesota Department of Natural Resources—Carol D. Hall.

west-central counties provide the basis of its status in Minnesota. Its distributional limits will become clearer after additional surveys and DNA analysis.

HABITAT

The Western Tiger Salamander adapts to a variety of habitats. Although typically not far from a water source, it may occupy dry prairie habitats and woodlands and breeds in wetlands, shallow lakes, and stock ponds.

LIFE HISTORY

Terrestrial adults emerge from overwintering burrows shortly after the ice is melted from wetland and pond margins. Movement to breeding sites is often triggered by early spring rains.

Eggs are laid singly in short rows or in small clusters within a gelatinous mass. Larvae are from 1 to 1.4 centimeters (3.5 to 5.5 inches) long at hatching and initially lack limbs (Hammerson 1999). Limbs develop after larvae attain roughly 2.5 centimeters

Western Tiger Salamander breeding habitat. Photograph by Tony Gamble.

(1.0 inch) in length (Tanner, Fisher, and Willis 1971). The larval tail fin extends up the back, nearly to the base of the neck.

Overwintering sites consist of burrows that extend below the frost line. Aquatic forms will survive in well-oxygenated wetlands and small lakes.

REMARKS

Molecular analysis conducted by Shaffer and McKnight (1996) provided data indicating that the Eastern and Western Tiger Salamanders should be regarded as distinct species. The Gray Tiger Salamander (*A. m. diaboli*), formerly a subspecies of *A. tigrinum,* has been documented in Douglas, Swift, and Yellow Medicine counties. Tissue analysis is being conducted to confirm the status of the Blotched Tiger Salamander (*A. m. melanostictum*) in Minnesota.

Eastern Tiger Salamander

Ambystoma tigrinum

DESCRIPTION

The Eastern Tiger Salamander is a large, relatively common terrestrial salamander that is widespread in Minnesota. Characteristic features of Eastern Tiger Salamanders include a stout body, broad head, and small eyes. The back and sides of this amphibian may be dark brown, dark gray, dark green, or black with numerous yellow-green to yellow-gold bars, blotches, or spots. Eastern Tiger Salamanders tend to become more blotched and less spotted with age. In Minnesota, some Eastern Tiger Salamanders are almost totally black. The belly surface is lighter, and belly spots are not as vivid as dorsal spots. There are 11 to 14 costal grooves. Adult Eastern Tiger Salamanders reach a total length of 18 to 33 centimeters (7 to 13 inches) (Conant and Collins 1998). The largest Minnesota specimen, from Douglas County, with a total length of 34.5 centimeters (13¾ inches) is the largest reported specimen for this species. Males have swollen cloacas during the breeding season. The larval form is yellowish green with small, dark spots and external gills attached to the base of the head. Larvae reach adult size before transformation.

Eastern Tiger Salamander, adult. Photograph by Allen Blake Sheldon.

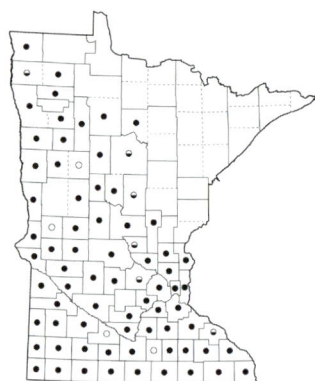

DISTRIBUTION

The Eastern Tiger Salamander ranges throughout the eastern half of the United States excluding the northeastern states. The salamander is found all across the state of Minnesota, although a number of counties lack verifiable records.

HABITAT

Eastern Tiger Salamanders utilize many types of habitats, including marshes, prairie ponds, farm ponds, woodland ponds, and lakes. Except in early spring and fall, adults spend most of their time underground in mammal burrows, crayfish burrows, or self-excavated burrows. Breckenridge (1944) observed Eastern Tiger Salamanders living in gopher and ground squirrel burrows.

LIFE HISTORY

Eastern Tiger Salamander eggs. Photograph by Minnesota Department of Natural Resources—Kelly Lynch Pharis.

Adults are commonly found aboveground during spring and fall migrations to and from breeding ponds and upland habitats. Many are accidentally trapped in window wells and swimming pools during migration. They are also encountered crossing

Eastern Tiger Salamander digging in gopher mound. Photograph by Minnesota Department of Natural Resources—Jeffrey B. LeClere.

highways during rainy weather, especially during the spring and fall. Overland movements are made nocturnally or on overcast days.

Eastern Tiger Salamanders are carnivorous during both larval and adult stages. Adults are known to eat earthworms, crickets, grasshoppers, other insects, small mice, and other amphibians, including their own larvae. Cannibalistic larvae develop grotesquely large heads (Hammerson 1982).

Warm, nocturnal spring rain stimulates movements to breeding ponds. Eighteen to over 100 eggs are laid in loose masses in early to late April after a unique strutting courtship (Kumpf 1934) during which the male nudges and pushes the female around the breeding pond and then leaves. The female then follows the male to reinitiate breeding. Eggs are attached to vegetation near the pond bottom. Transformation to the adult stage takes place during late August and September. The majority of salamanders found in the fall are newly transformed juveniles. They are sexually mature by their first spring.

The Eastern Tiger Salamander is capable of becoming sexually mature without transforming from a larva into a terrestrial form. This paedomorphic (or neotenic) form has been documented in Wisconsin (G. Casper, pers. comm.).

Skin secretions of Eastern Tiger Salamanders can be irritating to the eyes and mucous membranes of predators. This secretion serves as a useful defense mechanism.

REMARKS

Eastern Tiger Salamanders have been known to live 25 years in captivity (Bowler 1977). Adult Eastern Tiger Salamanders are often incorrectly identified as lizards. Salamanders superficially resemble lizards in body form, but lizards (reptiles) have scales, and salamanders (amphibians) do not. Larval Eastern Tiger Salamanders are used as fishing bait and sold as "waterdogs," sometimes misnamed as "mudpuppies."

The Eastern Tiger Salamander resembles the Spotted Salamander but has a broader head and spots that are highly variable in form and color and are distributed throughout its body. Because of difficulties in distinguishing it from the Western Tiger Salamander, DNA analysis may be the best option for confirming the geographic distribution of these two species in Minnesota. Previous county records of the former Tiger Salamander have been retained on distribution maps. As DNA analysis confirms the distribution of these two species, adjustments will be made to their range maps.

Family Plethodontidae— Lungless Salamanders

The plethodontids are the largest family of salamanders, with over 300 species worldwide. One-third of these species are found in the United States, where the greatest diversity occurs in the southern Appalachians. Two-thirds are neotropical, found in South and Central America, Europe, and recently Asia. These are the only salamanders that occur in tropical habitats.

The plethodontids are a highly variable family; however, they all share the same physical trait of being lungless. Respiration is accomplished through their skin, which needs to be moist to allow the transfer of oxygen and carbon dioxide. This is the only family that possesses a nasolabial groove that runs from the nostril to the upper margin of the lip. It aids in chemosensory functions such as searching for food and courtship. Most species, but not all, have four front toes, five back toes, and the ability to detach their tails as a defense. Their skin is adapted to moist environments and has antimicrobial compounds, which may serve it well as infectious diseases, including fungus, affect populations worldwide (Banning et al. 2008; Becker and Harris 2010).

Minnesota has two representatives of this family, the Four-toed Salamander and the Eastern Red-backed Salamander.

Four-toed Salamander

Hemidactylium scutatum

DESCRIPTION

This small salamander is the only terrestrial salamander in Minnesota with four toes on the front and hind feet. It is rusty brown on its dorsal surface and grayish along its sides, and the dorsum is white with bold black spots (Petranka 1998). They have 13 to 14 costal grooves. It has a constriction around the base of its tail where it can be detached to distract a predator.

Males have longer, squared snouts, while females' snouts are shorter and rounded (Petranka 1998). Adults measure from 5 to 10 centimeters (2 to 4 inches) in total length, and females average 15 percent longer in snout–vent length than males (Bishop 1941). Larvae have yellow-brown mottling and reach 11 to 15 millimeters (0.4 to 0.6 inch) (Petranka 1998).

DISTRIBUTION

Four-toed Salamander, adult. Photograph by Allen Blake Sheldon.

Four-toed Salamanders are found across the eastern United States, typically in small, disjunct colonies. Consequently, local

populations are vulnerable to catastrophic events and drastic habitat alterations. In Minnesota, they were first found in Itasca County in 1994 (Dorff 1995b). Since that time, they have been found in a number of north-central and northeast counties.

HABITAT

Four-toed Salamanders typically occupy mature upland deciduous or mixed deciduous-coniferous forest habitat. These salamanders occur most frequently in forests of glacial moraine landscapes where isolated wetlands are abundant. Breeding activity, egg deposition, and larval development occur in fishless wetland habitats within or adjacent to the forest. Upland forests provide cover, foraging sites, and overwintering habitat for juveniles and adults. The distribution and abundance of salamanders within the forest are influenced by temperature, moisture, and presence of woody debris for cover. Elimination or reduction of the forest canopy increases the temperature of the forest floor, reducing soil moisture and suitable cover.

LIFE HISTORY

Four-toed Salamanders breed in upland forest habitat in the fall. Males court females by circling about with their tail bent sharply upward at a right angle to their body. The female straddles the male's tail, and they walk in unison, with the male undulating

Four-toed Salamander belly. Photograph by Minnesota Department of Natural Resources—Carol D. Hall.

his tail from side to side (Petranka 1998). Males deposit spermatophores, which are picked up by the female using her cloaca.

In late April or early May, gravid females emerge from overwintering sites and move to nest sites just prior to egg laying. Wetlands utilized during the nesting season range from small shrub swamps to large conifer swamps, where females select a site near the forest-wetland interface. Sites selected for egg deposition typically consist of moss hummocks adjacent to open water, often at the base of alders or covering fallen logs. Females lay approximately 30 eggs and may remain with the egg cluster until time of hatching, giving off skin secretions that protect the eggs against the growth of fungus and bacteria (Banning et al. 2008). Multiple females may deposit eggs at the same site; some communal egg clusters have over 100 eggs, and often only one female remains with the clutches (Harris and Gill 1980). Banning et al. (2008) found that communal nesting in *H. scutatum* may facilitate the transmission of antifungal bacteria to embryos, giving them an advantage over nests with solitary clusters. Eggs are unpalatable to some invertebrates (Hess and Harris 2000). Upon hatching, larvae emerge from the moss and wiggle into adjacent shallow pools. Optimal water depth is at least 0.5 meters (1½ feet) at the time of hatching to provide adequate moisture necessary during the six weeks of larval development. Juveniles disperse into the adjacent forest two to three weeks after transformation.

Four-toed Salamanders are not freeze tolerant and escape freezing conditions by moving below the frost line in mammal burrows or root cavities created by rotting stumps (Blanchard 1933).

Four-toed Salamander, larva. Photograph by Allen Blake Sheldon.

REMARKS

The DNR lists the Four-toed Salamander as a species of special concern due to its unique habitat needs, low dispersal ability, and the threat of forest fragmentation (Minnesota DNR 2013). It is also considered a Species of Greatest Conservation Need (Minnesota DNR 2006).

Eastern Red-backed Salamander
Plethodon cinereus

DESCRIPTION

The Eastern Red-backed Salamander is appropriately named for its brick-red dorsal stripe, though some may have a dull brownish stripe. The sides and belly vary from brown to gray, with white flecking on the belly giving a salt-and-pepper appearance (Pfingsten and Downs 1989). A gray, or "leadback," phase and an all-red, or "erythristic," phase are also known, but they have not been documented in Minnesota. A large specimen is 10 centimeters (4 inches) in total length; the average ranges from 6 to 8 centimeters (2½ to 3 inches), and males are slightly smaller than females (Conant and Collins 1998). Snout–vent lengths of adults are greater than 3.7 centimeters (1.5 inches) (Sayler 1966).

The body is long and slender with four small, delicate legs. There are 18 to 19 distinct costal grooves. The Eastern Red-backed Salamander has five toes on the hind feet. The Four-toed Salamander is similar in its dorsal appearance to the Eastern Red-backed Salamander, but it has four toes on the hind feet. The Four-toed Salamander also has a bright-white belly with bold black flecks. Eggs are visible through the translucent ventral skin of gravid females.

Red-backed Salamander, adult. Photograph by Allen Blake Sheldon.

DISTRIBUTION

The Eastern Red-backed Salamander is found in the northeastern United States and southeastern Canada from southern Quebec and Ontario to North Carolina and Indiana.

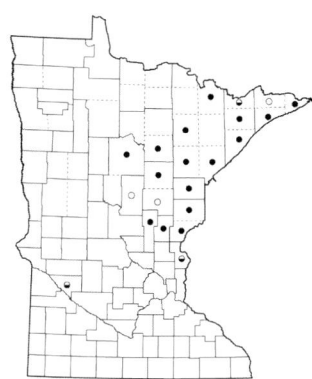

Most of the Eastern Red-backed Salamander records in Minnesota are from the northeast region of the state, from Cass County east into the northeast corner. There is one disjunct record from along the Minnesota River near Montevideo in Chippewa County, but this population is no longer extant.

HABITAT

Eastern Red-backed Salamanders are completely terrestrial salamanders closely associated with forested habitats. Breckenridge (1944) reported that specimens collected in northeastern Minnesota are commonly associated with pockets of hardwoods. In deciduous forests the salamanders are found under decaying logs, rocks, and leaf litter. The Chippewa County record is from a floodplain forest along the Minnesota River that extends into former prairie habitat (Breckenridge 1944). They are also found in coniferous forest habitat. Eastern Red-backed Salamanders overwinter in animal tunnels up to 1 meter (3 feet) below the surface (R. Caldwell 1975).

LIFE HISTORY

Female Eastern Red-backed Salamanders reach maturity in their second or third summer (Sayler 1966; Nagel 1977). They deposit their eggs in damp substrates such as in rotten logs or in soil under rocks and logs. Yurewicz and Wilbur (2004) found females laid an average of 7 eggs, although clutch sizes ranged from 2 to 12. They are the only Minnesota salamander that does not have an aquatic larval stage. The female guards her eggs and the young for up to three weeks after hatching (Pfingsten and Downs 1989). Juveniles look like miniature versions of the adults.

The Eastern Red-backed Salamander spends its life in a very humid environment. They are rarely found when the humidity is below 85 percent (Heatwole 1962). Seasonal variations in humidity are correlated with salamander activity. In the spring and fall, when damp conditions normally occur, adults are more

Red-backed Salamander eggs. Photograph by Allen Blake Sheldon.

easily found. The Eastern Red-backed Salamander has a small home range of less than 25 square meters (270 square feet) (Kleeberger and Werner 1982).

Aggressive behavior can result in tail loss, which reduces body fat reserves, or in scarring of the nasolabial groove, which may reduce the salamander's ability to gather chemical cues from its environment (C. W. Brown 1968).

When Eastern Red-backed Salamanders are active, they eat a wide variety of small invertebrates including worms, sow bugs, centipedes, and spiders (Bishop 1941). In Indiana during December, Caldwell (1975) found Eastern Red-backed Salamanders overwintering in ant mounds had stomachs full of ants. Eastern Red-backed Salamanders are preyed on by small snakes, shrews, and larger salamanders.

REMARKS

Burton and Likens (1975) found the biomass of salamanders to be twice that of birds in a New Hampshire study site. Eastern Red-backed Salamanders made up 94 percent of the salamander biomass and numbered 2,950 salamanders per hectare (1,200 per acre). The abundance of these small vertebrates in the eastern portion of its range indicates their importance in local ecosystem function.

The Red-backed Salamander is considered a Species of Greatest Conservation Need by the Minnesota DNR (2006).

Family Proteidae— Waterdogs and Mudpuppies

Waterdogs are a small family of six species worldwide. Five species are found in the United States and southern Canada, and one species is found in southern Europe. The salamanders in this family are all neotenic. As adults, they possess both gills and lungs. They never leave the water, even under drought conditions.

Minnesota has one species, the Mudpuppy, which belongs to the only North America genus.

Mudpuppy

Necturus maculosus

DESCRIPTION

The Mudpuppy is the only fully aquatic salamander in Minnesota. An adult Mudpuppy is a brownish color with dark blotches over the entire body. Juveniles are dark brown with two broad, lateral yellow stripes. The deep-red gills are the most striking feature. Mudpuppies have a large, squarish head with small eyes and nostrils, and their front and hind legs have four toes. The tail is laterally flattened with a dorsal and ventral fin. The Mudpuppy is one of the largest salamanders in Minnesota with an average length of 33 centimeters (13 inches). Some specimens have been recorded up to 40 centimeters (16 inches) (Bishop 1941). The sexes are similar, but males have a crescent-shaped groove on the front of the vent (Bishop 1941) and a swollen cloaca during the breeding season. Larval and paedomorphic Tiger Salamanders are sometimes confused with Mudpuppies, but they have five toes on their hind legs and a rounder head, and their tail fin extends onto the trunk of the body, almost all the way to the head.

Mudpuppy, adult. Photograph by Allen Blake Sheldon.

DISTRIBUTION

Mudpuppies are found in the east-central United States from the Great Lakes south to northern Alabama. They are found west of the Appalachian Mountains through Missouri and north to Minnesota.

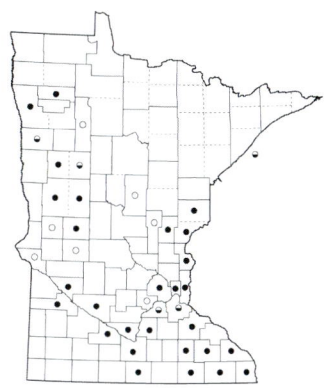

In Minnesota, Mudpuppies are primarily restricted to the Minnesota, Red, St. Croix, and lower Mississippi River drainages. St. Anthony Falls acts as a barrier to the upper Mississippi River (Cochran 1991).

HABITAT

The Mudpuppy is found in large to medium rivers and large lakes. In Minnesota, Mudpuppy habitats range from swift gravel-bottom streams to slow muddy rivers. Rivers with high turbidity are used if silt-free gravel areas are present for nesting (Pfingsten and Downs 1989). Mudpuppies require cover for breeding and may use rocks, sunken logs, or submerged, abandoned tires. Young Mudpuppies stay in shallow water or riffles, while adults use deeper water (Pfingsten and Downs 1989). They have

Close-up of Mudpuppy. Photograph by Allen Blake Sheldon.

been documented 27 meters (88.5 feet) deep in Lake Michigan (Reigle 1967).

LIFE HISTORY

Adult Mudpuppies spend the entire year in the same habitat, and they are active year-round. The DNR regularly receives reports of Mudpuppies caught by anglers ice fishing on the St. Croix River.

Mudpuppies breed in the fall and early winter, although spring breeding is also suspected to occur (Bishop 1941). Their courtship ritual is similar to other salamanders (Bishop 1941); males deposit large spermatophores under rock slabs or logs. Females store the sperm over the winter, delaying egg fertilization until spring (Harris 1959). Most females spawn within a two- to three-week period (Bishop 1926), laying their eggs in late spring or summer under rocks or logs in silt-free areas and guarding

Mudpuppy eggs. Photograph by Allen Blake Sheldon.

Mudpuppy, juvenile. Photograph by James E. Gerholdt.

them during development. The eggs, laid in groups of 50 to 100, take one to two months to hatch depending on water temperature. Warmer streams and lakes allow faster development.

Size at hatching is 22 millimeters (⅞ inch) (Bishop 1941). Young Mudpuppies are very secretive, and there are very few records of juveniles from Minnesota (Cochran, pers. comm.). The color pattern changes during the five years it takes to reach maturity; dark dorsal bands and yellow stripes change to dark blotches throughout the body. When Mudpuppies reach approximately 13 to 15 centimeters total length, their markings become similar to adults' (Bishop 1941).

Mudpuppies are carnivorous and eat nearly anything that fits into their mouth, including crustaceans, insects, worms, fish, and even other salamanders (Bishop 1941). A sample of 340 Mudpuppies from New York revealed that crayfish and aquatic insects comprised the major volume of food items (Hamilton 1932). Mudpuppies feed year-round.

REMARKS

The Mudpuppy was once considered vile and poisonous. Local fishermen would cut them off their lines because they erroneously believed the Mudpuppy had dangerous spines and bites (Breckenridge 1944). Adult Mudpuppies have few natural enemies and can live to be over 30 years old (Bonin et al. 1995). Lampricide may have eliminated Mudpuppy populations adjacent to Lake Superior, during efforts to control the sea lamprey (Matson 1990).

Mudpuppies are the only known host of the salamander mussel, a threatened species in Minnesota.

The subspecies found in Minnesota is the Common Mudpuppy *(N. m. maculosus).* It is listed as a species of special concern in Minnesota and considered a Species of Greatest Conservation Need by the Minnesota DNR (2006, 2013). Although not currently regulated by the Minnesota DNR, Mudpuppies have been harvested for sale to biological supply companies.

Family Salamandridae—Newts

Newts are found in Europe, Asia, North Africa, and North America. The United States has only 6 species in two genera of the 70 species found across the family's range. Three species (genus *Notophthalmus*) are found in the eastern United States, and 3 species along the West Coast (genus *Taricha*). Most species in the family are found in Europe, with representatives in northern Africa, China, Southeast Asia, and Japan.

Terrestrial forms of these salamanders have rough skin, while aquatic forms have smooth skin. Costal grooves are absent in both forms. Breeding is variable; some species breed in the water, and some on land. Subadult newts may spend several years as terrestrial salamanders, called efts.

Minnesota has only one species, the Eastern Newt, belonging to the newt family.

Eastern Newt
Notophthalmus viridescens

DESCRIPTION

The Eastern Newt is a small salamander with a total length of 6.5 to 14 centimeters (2½ to 5½ inches) (Conant and Collins 1998). Adults are olive green with a row of reddish spots on each side. The ventral surface is yellowish with scattered black dots. Aquatic adults have a vertically flattened and finned tail, which assists in swimming. The rear legs are larger than the front legs. Aquatic forms have smooth skin. Males can be distinguished by their enlarged tail fins and swollen vents during the breeding season.

The juvenile, terrestrial phase (eft) is smaller than the adult, having a length of 4 to 8 centimeters (1½ to 3 inches). Individuals are orange to reddish brown with a lighter belly. Efts found south and east of Minnesota tend to be bright orange. Eft skin appears dry and granular.

Eastern Newt, adult. Photograph by Barney Oldfield.

DISTRIBUTION

The Eastern Newt is found throughout most of the eastern United States, west to eastern Texas in the south and to Minnesota in the north. In Minnesota the known range is spotty. There are scattered populations in the upper Mississippi River drainage, and they appear to be more common in the northern counties.

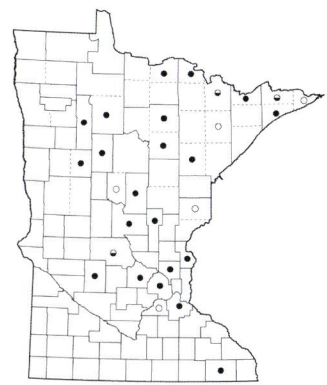

HABITAT

In Minnesota Eastern Newts are found in temporary and permanent lakes, streams, and wetlands adjacent to woodland habitats. They can occur in coniferous and deciduous forests as long as there is a good understory with numerous decaying logs and abundant leaf litter.

LIFE HISTORY

Efts are encountered on land in summer and fall, especially during rainy or overcast days with high humidity. The juvenile stage may last four to seven years or be nonexistent (Breckenridge 1944; Gill 1978). In Minnesota it appears that the eft stage may be short (i.e., one or two years), but better documentation is needed.

Eastern Newt efts, color variation.
Photograph by Jeffrey B. LeClere.

Aquatic adults can be found in streams and ponds year-round. In the fall they migrate to deeper ponds because shallow ponds become oxygen deficient in the winter (Pfingsten and Downs 1989). Adults overwinter on land, as do the efts, if conditions are right (Healy 1974). During the summer, sedentary adults rarely move from pond to pond, but efts will wander wherever there is adequate habitat (Healy 1975).

Aquatic adults eat various aquatic insect larvae, amphibian eggs and larvae, and fingernail clams. Efts eat terrestrial insects, small crustaceans, and worms.

Adult Eastern Newts breed in the spring in fish-free streams and ponds. In Minnesota breeding probably occurs in early May. After congregating at breeding sites, Eastern Newts conduct a courtship dance in which the male fans the female with his tail (Verrell 1982). The male deposits a spermatophore on the bottom of the pool, and the female then positions her cloaca over the spermatophore to pick it up. Two to three days later the female lays 6 to 10 eggs (Pfingsten and Downs 1989). The female may breed up to 30 times over the course of the breeding season (Gill 1978). Eggs hatch in two to four weeks depending on the temperature of the water.

The larvae spend the summer in the natal pond feeding on small aquatic insects and growing to a length of 4 centimeters

Eastern Newt, larva. Photograph by Allen Blake Sheldon.

Eastern Newt, eft. Photograph by Allen Blake Sheldon.

(1½ inches). They are preyed on by other salamander larvae, adult newts, and large aquatic insects. By late summer larvae usually transform into terrestrial efts, or they may skip the eft phase and transform directly into aquatic adults.

The skin of an eft is toxic and makes most predators ill if they eat them (Bishop 1941). The eft's orange-red coloration serves as a warning signal to potential predators. Efts are often active during the day and wander boldly across the forest floor, protected by their warning coloration.

REMARKS

The subspecies of Eastern Newt found in Minnesota is the Central Newt, *Notophthalmus viridescens louisianensis* (Conant and Collins 1998).

Class Reptilia

Class Reptilia

Of all living and extinct animals, dinosaurs have captured the imagination of humans more than any other group. Highly successful, dinosaurs ruled during the Age of Reptiles, which lasted for at least 200 million years. The earliest known reptile fossil is 315 million years old, and the last known large dinosaurs perished 65 million years ago. Reptiles have continued to the present day with over 9,000 known living species.

Four orders of reptiles are found on the earth today. Order Rhynchocephalia contains one species, the Tuatara, found on several New Zealand islands. It is the sole survivor of a large group of primitive reptiles. Twenty-four species of the order Crocodilia (crocodilians) are found throughout the world, of which 2 are found in the United States, primarily in the tropics and subtropics. Order Testudines (turtles and tortoises) is comprised of distinctive reptiles with a protective shell. This order includes 326 species worldwide, of which 59 are found in the United States, including 11 species in Minnesota. Order Squamata (lizards and snakes) is divided into two suborders. Lacertilia (lizards), the largest suborder, comprises 5,520 species worldwide, of which 120 are found in the United States, including 3 species in Minnesota. Finally, a relatively modern group of reptiles, Serpentes (snakes), also have a worldwide distribution. This group is made up of 3,220 species, of which 155 are found in the United States, including 17 species in Minnesota.

Turtles have an upper shell (carapace) and lower shell (plastron) connected at the sides by a bridge, allowing the reptile to withdraw its head, legs, and tail inside for protection. Softshells have pliable shells, in contrast to the rigid bony structure of other turtle species. Turtles have claws on their toes and lack teeth. Most lizards have four legs, and claws on their toes, although some species lack limbs. The tails of many lizard species are easily broken off when grasped by a predator, and the shortened tail is generally capable of regrowing. Lizards have eyelids and ear openings on the sides of their heads. Snakes are legless. They

lack ear openings and have forked tongues. Instead of eyelids, snakes have a clear scale covering the eye, called a spectacle. A segmented rattle is found on the end of the tail of our native rattlesnakes.

The most distinguishing feature of reptiles is their skin, which is comprised of scales or plates. In comparison to amphibians, the skin of reptiles markedly retards water loss and provides greater protection against abrasion. Reptiles are less dependent on water; thus, many species have been able to adapt well to dry terrestrial habitats.

Brightly colored skin pigments can be found in many species. Skin color plays important roles in attracting mates, eluding predators, and capturing prey. Several species of lizards are capable of remarkable changes in skin color. To make room for growth and to replace old with new, reptiles periodically shed their skin. Starting at the nose, snakes crawl out of their old skin as it turns inside out. Lizards shed in patches, and most turtles periodically lose the outer covering of their scutes. Young and growing animals shed more frequently than adults.

Reptiles do not undergo metamorphosis, unlike amphibians. A newborn reptile is a miniature replica of its parents and generally closely resembles them. They breathe air with lungs throughout life.

With only a few exceptions, reptile reproduction is bisexual with both sexes contributing genetic material to the next generation. Courtship and breeding behaviors can be simple or elaborate, and they vary greatly from species to species. In some cases these courtship activities are instrumental in preventing interspecific breeding. All reptiles have internal fertilization. Females of several species have the capacity to store live sperm in their reproductive tracts for extended periods, and egg fertilization can occur up to several years after breeding. Male lizards and snakes have paired copulatory organs called hemipenes. However, during copulation only one of the organs is used. Refer to individual species accounts for more detailed descriptions of reproductive behavior.

The shelled amniote egg was an important evolutionary step for reptiles because it broke their tie to water for egg development. They became free to reproduce in terrestrial environments. Many reptile species lay eggs; however, a number of lizards and snakes give birth to fully developed young. Eggs are generally laid by the female in moist soil or humus and abandoned during development. Several species of reptiles demonstrate egg tending and provide primitive parental care to their young. Female skinks found in Minnesota brood their egg clutch until the young hatch. Newborn rattlesnakes stay with their mother for a period of 10 to 14 days after birth.

Many reptiles have a highly developed sense of smell, which is used primarily to procure food and locate mates. Lizards and snakes use their tongues to "taste" the environment. Their tongues relay small particles of their surroundings to a pair of Jacobson's organs, located in the back of the mouth, for sensory analysis. Sight is well developed in many species. Aquatic turtles depend on visual cues to detect approaching predators while they bask. Reptilian vocalizations are rare, and their sense of hearing is poorly developed. Snakes are totally deaf, but they are remarkably sensitive to ground vibrations. Pit vipers possess a highly developed pair of infrared-sensitive facial pits. The pits are capable of detecting very small changes in temperature (i.e., the snake can tell when a warm-blooded prey animal comes within striking distance).

Food requirements vary among reptilian species. Turtles may be herbivorous, carnivorous, or omnivorous. Hatchlings of Minnesota's turtle species are carnivorous, but they may begin eating vegetation as they mature. A large percentage of lizards are carnivorous; all snakes and crocodilians are carnivorous. Snakes consume their prey whole, and elastic articulations in their jaws make them capable of swallowing food items larger than their head. Three methods of securing prey are used by snakes in Minnesota. The seize-and-swallow method is used by such species as the North American Racer and gartersnakes. Constrictors, such as the Gophersnake and Western Foxsnake, grasp the animal in their mouth and constrict it with body coils. Death of the prey is due to suffocation. Rattlesnakes use venom to immobilize their prey before they swallow it.

Some reptilian species mature in less than 12 months, while others require as many as 20 years to reach adulthood. Longevity is known to exceed 100 years in some species of turtles and tortoises, although 30 to 75 years is average for most Minnesota turtles. Large snakes can live up to 20 years, while the life span of small snakes and the majority of lizards does not exceed 10 years.

Family Scincidae—Skinks

Scincidae is the largest lizard family with nearly 1,426 species worldwide. Representatives of 137 genera are found on every continent except Antarctica, and they are especially abundant in the tropics of Southeast Asia. The United States has 16 species belonging to 3 genera.

A typical skink has numerous shiny, smooth scales covering a sausage-shaped body and tail. The tail readily breaks off when grasped by a predator, and the wriggling tail segment serves as a decoy while the skink runs for safety. These diurnal lizards are chiefly terrestrial and customarily feed on small invertebrates. Virtually all New World species lay eggs, and females typically guard the egg clutch until hatching.

Minnesota has two species of skinks, both belonging to the genus *Plestiodon.* The Prairie Skink is relatively common across the state, but the Common Five-lined Skink is found only in selected localities.

Common Five-lined Skink
Plestiodon fasciatus

DESCRIPTION

The Common Five-lined Skink is a small, robust lizard with smooth, shiny scales and relatively small legs. Coloration varies with age and breeding condition, but the lizard gets its name from five yellowish stripes that extend longitudinally along the back and sides. The middorsal stripe forks at the neck forming a narrow V across the top of the head, which becomes muted or absent in older adults. The black or dark-brown background coloration of juveniles and young adults creates a distinct five-lined pattern that markedly contrasts with the bright metallic-blue tail. Mature females retain a lined pattern on a lighter background, whereas mature males may become a uniform gray or light brown. Tails of adult males are generally gray, while those of females tend to be blue gray. The nose, lips, cheeks, and throat of a male in breeding condition are bright orange red.

The total length of the Common Five-lined Skink ranges from 12.5 to 21.5 centimeters (5 to 8½ inches), and the maximum

Common Five-lined Skink, adult male. Photograph by Allen Blake Sheldon.

snout–vent length is 8.6 centimeters (3⅜ inches) (Conant and Collins 1998).

The Common Five-lined Skink is very similar in appearance to the Prairie Skink; a close look is needed to make the distinction. Prairie Skinks have dark-brown bands along their sides and broad, tan stripes down their back. Prairie Skinks do not have the V on their head. In contrast to the Common Five-lined Skink, the Six-lined Racerunner has dull, rough scales rather than shiny, smooth ones, and it has large rectangular belly plates in contrast to the small, uniform abdominal scales of skinks.

DISTRIBUTION

The Common Five-lined Skink ranges across the eastern half of the United States excluding the extreme northeastern states and southern Florida. This species has an unusual, patchy distribution in Minnesota. Breckenridge (1944) reported records from Redwood and Yellow Medicine Counties along the Minnesota River. Further fieldwork in the 1980s confirmed the earlier localities and documented additional records in Renville, Fillmore, and Houston Counties (Lang 1982; Moriarty 1986). Five-lined Skinks were confirmed in Chisago County in 1992. These scattered localities in Minnesota are disjunct from the primary range of this lizard.

HABITAT

The habitat for the Common Five-lined Skink is generally described as humid woodlands and wooded lots with decaying leaf litter, stumps, and logs (Conant and Collins 1998). In Minnesota, however, the species is found on or near granite outcrops in the dissected terrain of the Minnesota River valley (Lang 1982). This species is found in southeastern Minnesota in association with limestone outcrops and bluff prairies in proximity to deciduous forest. Common Five-lined Skinks retreat to depths of 1.5 to 3 meters (5 to 10 feet) in rock fissures and cracks below the frost line to survive Minnesota winters (Lang 1982).

Common Five-lined Skink, juvenile. Photograph by Allen Blake Sheldon.

LIFE HISTORY

In Minnesota, the diurnal Common Five-lined Skink becomes active in early May at air temperatures of 15.5° to 32°C (60° to 90°F). After a flurry of activity in late spring and early summer, this species becomes increasingly difficult to find before it retreats to hibernation sites by September. This lizard is secretive and remains under protective cover much of the time. It can be found under rocks and ground debris such as cardboard, sheets of corrugated metal, tar paper, and cloth (Lang 1982). Although on occasion they climb into low shrubs to bask or forage, they are primarily terrestrial. According to Fitch (1954), their home territory has a diameter of 9 to 27.5 meters (30 to 90 feet).

These lizards feed on small invertebrates such as crickets, locusts, beetles, insect larvae, caterpillars, grasshoppers, moths, and snails. The predominant food for Common Five-lined Skinks in Minnesota is roaches and spiders. They lap droplets of dew from vegetation for their water needs (Lang 1982).

Breeding activity peaks during the third and fourth weeks of May in Minnesota. Common Five-lined Skinks emerging from hibernation in their second year are sexually mature. Males aggressively defend territories against intruders, and these interactions occasionally result in fighting. Males locate females by sight and scent. Upon finding a receptive mate, the male grasps the skin behind her head with his mouth while curling his tail

Adult female Common Five-lined Skink brooding eggs. Photograph by Allen Blake Sheldon.

Common Five-lined Skink habitat in southeast Minnesota. Photograph by Minnesota Department of Natural Resources—Carol D. Hall.

under hers in order to achieve copulation. About a month later, the gravid female excavates a small chamber under a rock or within decaying vegetation and deposits 5 to 13 (average of 9) pliable, white eggs. She stays with the eggs until they hatch, guarding against predators and keeping the nest free of spoiled eggs by eating them. The eggs require 30 to 60 days to hatch. Hatchlings vary from 5 to 6.4 centimeters (2 to 2½ inches) in total length (Conant and Collins 1998).

If a predator encounters a Common Five-lined Skink and grasps its tail, the tail readily disjoints, leaving a wriggling decoy while the skink makes a dash for safety. When frightened and cornered, a Common Five-lined Skink can break off its own tail by pushing it against a firm surface. Very little blood is lost. The tail regenerates over time, but the replacement appendage is never as long or colorful as the original. Common Five-lined Skinks inevitably attempt to bite a human captor, but ordinarily their small teeth cannot penetrate skin.

Natural predators of this species include snakes, hawks, and small mammals such as shrews, raccoons, skunks, and opossums. Domesticated cats may also catch and eat these lizards.

REMARKS

In Oldfield and Moriarty (1994) the Common Five-lined Skink was called the Five-lined Skink with the scientific name of *Eumeces fasciatus*. The word *common* was added to the English name to remove confusion with the Southeastern Five-lined Skink (Crother 2012). Recent evidence showed the large genus *Eumeces* contained multiple genera. Members of the group in the United States were put in the genus *Plestiodon* (Brandley, Schmitz, and Reeder 2005; Griffith, Ngo, and Murphy 2000).

The bright-blue tail of young Common Five-lined Skinks is the basis for names such as "blue-tailed skink" and "blue devil."

In Minnesota the Common Five-lined Skink is classified as special concern and listed as a Species of Greatest Conservation Need (Minnesota DNR 2006, 2013).

Prairie Skink

Plestiodon septentrionalis

DESCRIPTION

Of the three lizard species found in Minnesota, the Prairie Skink is the most widespread. This medium-sized skink has small legs and a body that is cylindrically shaped in cross section. Body scales are smooth and shiny. Three wide, light-tan stripes separated by two narrow, dark lines run longitudinally from the head onto the tail. Additionally, three black or dark-brown stripes are found on each side separated by two narrow, white stripes. Coloration of the tail is essentially the same as the body, although the stripes become less defined toward the tip. The abdomen is unmarked gray or tan, and the lizard's legs are light brown. During the breeding season, adult males have bright-orange throats, lips, and chins. Hatchling Prairie Skinks are black with seven thin, yellowish stripes and a brilliant metallic-blue tail.

Prairie Skinks range in total length from 13.3 to 22.4 centimeters (5¼ to 8¾ inches). The maximum body length (snout–vent) is 9 centimeters (3½ inches) (Conant and Collins 1998).

Prairie Skink, adult female. Photograph by James E. Gerholdt.

The two species of Minnesota skinks look very similar. Prairie Skinks, in contrast to Common Five-lined Skinks, lack an inverted V on top of their head. Prairie Skinks have wide, light-brown stripes down their back and wide, dark-brown or black stripes on their sides, while Common Five-lined Skinks have narrow, light-colored stripes along their back and sides.

DISTRIBUTION

The Prairie Skink is found in a north-to-south band from Minnesota and western Wisconsin to eastern Texas. An isolated population is found in southern Manitoba. With the exception of the Arrowhead region of the northeast and the bluff lands of the southeast, the Prairie Skink is found in suitable habitat across Minnesota.

HABITAT

Open, grassy areas in association with pine barrens and oak savannas with loose sandy soil and scattered rocks are ideal Prairie Skink habitat. They are also found on gravelly glacial outwashes and extensive rock outcrops if there is sufficient vegetation to

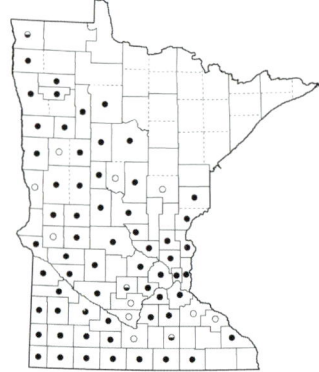

Prairie Skink, adult male. Photograph by James E. Gerholdt.

provide cover and insect food. Vogt (1981) reports that they frequently live on sandbanks along creeks and rivers. They are found in urban areas with adequate habitat. Prairie Skinks hibernate below the frost line in self-constructed burrows at depths of 0.3 to 1.4 meters (1 to 4½ feet), either singly or in small groups (Breckenridge 1943).

LIFE HISTORY

Prairie Skinks typically begin their activity in early May. They remain active through September, but adults become difficult to find late in the season. Highly secretive, this species spends much of the day under rocks and other ground cover (Collins 1982). They remain underground when air temperatures are too cool for activity, below approximately 60°F (16.5°C) (Fitch 1954).

Prairie Skinks feed principally on small arthropods including crickets, grasshoppers, treehoppers, leafhoppers, beetles, caterpillars, and spiders. They occasionally consume the young of their own kind (Breckenridge 1943).

Courtship has not been described from Minnesota, but ordinarily the breeding season is the last half of May, when males sport their bright breeding colors. The female digs a nest under a rock or log and lays 5 to 13 eggs (average of 9) in late June or early July. The eggs are dirty white in color and have soft leathery shells. The average egg size at the time of laying is 0.8 by 1.3 centimeters (⁵⁄₁₆ by ½ inch), and the egg increases to 1.1 by 1.9 centimeters (⁷⁄₁₆ by ¾ inch) just before hatching (Breckenridge 1943). The egg development period ranges from 40 to 52 days depending on the weather. The female skink guards and tends the eggs until they hatch, and she provides 2 to 3 days of maternal care to the hatchlings before they leave the nest (Somma 1987). Young Prairie Skinks grow rapidly and are ready to begin breeding during their third season or when they are just shy of two years of age (Breckenridge 1943).

Like the Common Five-lined Skink and many other lizard

Prairie Skink nest. Photograph by Minnesota Department of Natural Resources—Carol D. Hall.

Prairie Skink, hatchling.
Photograph by Allen Blake
Sheldon.

species, this reptile uses an easily disjointed tail to help baffle predators. The tail regrows, but it never attains its original length or color. This species generally attempts to bite when handled, but its small mouth and teeth are unlikely to break the skin of a human hand.

Natural predators include northern harriers, barred owls, kestrels, shrikes, striped ground squirrels, and raccoons (Vogt 1981). Prairie Skinks impaled on barbed wire by loggerhead shrikes have been found in Clay County (Oldfield and Moriarty 1994). The Plains Hog-nosed Snake may be an important predator of the Prairie Skink in Minnesota where the two species share the same habitat (Oldfield and Moriarty 1994).

REMARKS

In Oldfield and Moriarty (1994) the scientific name of the Prairie Skink was *Eumeces septentrionalis*. Recent evidence showed the large genus *Eumeces* contained multiple genera. Members of the group in the United States were put in the genus *Plestiodon* (Brandley, Schmitz, and Reeder 2005; Griffith, Ngo, and Murphy 2000).

According to Conant and Collins (1998), the subspecies of Prairie Skink found in Minnesota is the Northern Prairie Skink *(Plestiodon septentrionalis septentrionalis)*. In the past this species was referred to as the Black-banded Skink (Breckenridge 1943).

Family Teiidae—
Racerunners and Whiptails

Teiidae is a large family of 123 species and 10 genera of New World lizards, of which the vast majority of species are found in South America. They are built for speed with well-developed legs, a pointed head, and a long, thin tail. Small granular scales on the back contrast with large rectangular belly scales. Their tongue is long and deeply forked like that of snakes. The smallest teiid is 7.6 centimeters (3 inches) long, and the largest exceeds 122 centimeters (48 inches). Racerunners and whiptails are diurnal, terrestrial, and oviparous. Several species of teiids are unisexual and reproduce by parthenogenesis.

There are 22 species of racerunners and whiptails native to the United States. All species belong to the genus *Aspidoscelis*. One teiid ranges into Minnesota, the Six-lined Racerunner.

Six-lined Racerunner

Aspidoscelis sexlineata

DESCRIPTION

The Six-lined Racerunner is an appropriate name for this lizard. This species is very capable of running and winning a race with a would-be human captor. An indefinite tan middorsal stripe originating at the base of the head is flanked on either side by three distinct yellow or yellowish-green stripes that extend to the base of the tail. The top of the head, the dorsal surface of the legs, and the tail are light brown. The area on the back between the stripes is dark brown. The belly of females and juveniles is light gray or white, and the belly of adult males is bluish. During the breeding season males develop a light-blue coloration on their chin and lips, and they have a brilliant lime-green wash along the sides of their head and the anterior third of their body. The tails of hatchling and juvenile Six-lined Racerunners are light blue, and the body stripes extend onto the tail for about a third of its length.

Six-lined Racerunner, adult female. Photograph by James E. Gerholdt.

The Six-lined Racerunner reaches a total length of 24 centimeters (9½ inches) and has a snout–vent length of 6 to 8 centimeters (2⅓ to 3⅛ in; Vogt 1981).

In contrast to skinks, members of the genus *Aspidoscelis* have pointed noses, large scales on top of their head, and rectangular belly plates.

DISTRIBUTION

Six-lined Racerunners range across the southeastern two-thirds of the United States with a northern extension along the Mississippi River valley into Minnesota and Wisconsin. In Minnesota, they are found in the southeast portion of the state.

HABITAT

In Minnesota, Six-lined Racerunners are found in open, sandy or gravelly areas with sparse ground vegetation. Prairies on south-facing hills and sand outwashes in river floodplains are preferred, but populations are also encountered along rock and cinder fills of railroad tracks and man-made dikes along the Mississippi River (Oldfield and Moriarty 1994). These lizards are often found in colonies of sizable populations. Hibernation and estivation take place in underground burrows. These refuges are ordinarily self-excavated in loose soil.

LIFE HISTORY

The Six-lined Racerunner is the least cold tolerant of all Minnesota reptiles. Adults emerge from overwintering in mid-May, and by late August they have retired for the season. Hatchlings remain active and continue feeding into September to increase body fat stores (Vogt 1981).

Fitch (1958a) found that on warm, sunny days Six-lined Racerunners are active from 8:00 a.m. until 3:00 p.m. and that the optimal air temperature for activity is 34°C (93°F). Adults have been found active at an air temperature of 22°C (72°F) on steep bluff prairies in Winona County in mid-May (Oldfield and Moriarty 1994). They may seek shelter under vegetation or underground for brief periods during the hottest part of the day. On cool days they remain in burrows or hidden under ground cover. Vogt (1981) determined that individuals maintain home ranges of 160 square meters (191 square yards) in Wisconsin, although Fitch (1958b) estimated home ranges to be 1,000 square meters (1,196 square yards or 0.25 acre) in eastern Kansas. Territorial behavior has not been observed with this species.

Six-lined Racerunners pursue and consume a variety of arthropod prey including grasshoppers, crickets, katydids, moths, beetles, bugs, ants, spiders, and flies. Before swallowing large grasshoppers, the lizard grasps and holds them in its jaws while scraping the prey against the ground to tear off the long rear legs. A keen sense of smell and good eyesight are used to find and follow quarry.

Courtship and breeding occur during late May and early June shortly after the lizards emerge from their winter hibernation. The male courts the female by displaying his vivid coloration. If the female is receptive, she allows him to grasp the skin

Six-lined Racerunner, hatchling. Photograph by Allen Blake Sheldon.

Six-lined Racerunner in burrow. Photograph by Minnesota Department of Natural Resources—Carol D. Hall.

on the back of her neck with his jaws while he positions his tail under hers to achieve copulation.

Clutches of one to six eggs are laid in a burrow dug by the female at a depth of 10 centimeters (4 inches) in loose sand during mid-June. Egg tending does not occur. In Missouri and farther south older females normally lay two clutches each season (Johnson 1987). The white eggs with thin, leathery shells average 1.6 by 0.9 centimeters (⅝ by ⅜ inch), and they normally take two months to hatch (Johnson 1987). Hatchlings have a total length of 5 centimeters (2 inches).

Six-lined Racerunners depend on speed for self-defense. They move with nervous short spurts when foraging, and at the slightest hint of danger they quickly accelerate to speeds of 29 kilometers per hour (18 miles per hour; Vogt 1981). The first glimpse of these lizards is generally a blurred streak as they make a dash to hide in ground vegetation or a burrow. They may lose their tail to predators, but it is not as fragile as a skink's tail as evidenced by the fact that most Six-lined Racerunners encountered in the wild still possess their original tail. They have been known to self-break their tail without being grasped (LeClere 2013). This behavior distracts a predator and allows the lizard to escape. Natural predators include birds of prey, small mammals, and snakes. Racers are significant predators because they have the speed to pursue and capture their prey. Milksnakes feed on Six-lined Racerunners when the lizards are less active in their burrows.

REMARKS

In Oldfield and Moriarty (1994) the scientific name of Six-lined Racerunner was *Cnemidophorus sexlineatus*. Reeder, Dessauer, and Coles (2002) split the original genus and placed all the North American species in *Aspidoscelis*.

The Six-lined Racerunner has been listed as a Species of Greatest Conservation Need by the Minnesota DNR (2006).

The Prairie Racerunner, *Aspidoscelis sexlineata viridis,* is the subspecies found in Minnesota (Conant and Collins 1998).

Two colorful and fitting nicknames for these lizards are "field-streaks" and "sandlappers."

Family Colubridae—
Colubrid Snakes

Colubridae is the largest family of snakes and contains over 652 species worldwide. Representatives of this interesting and diverse group of snakes are found on every continent except Antarctica. The United States has 65 species of colubrids.

Fifteen of Minnesota's 17 snake species belong to the colubrid family. They range in size from small Ring-necked and Red-bellied Snakes to large Gophersnakes and Western Ratsnakes. They may be stout bodied, such as Eastern Hog-nosed and Plains Hog-nosed Snakes, or slender and sleek, such as North American Racers and Smooth Greensnakes. Some species are live bearers, while others lay eggs. A few species produce a very mild toxin to help immobilize prey, but none are dangerous to humans. Other Minnesota colubrids are the Western Foxsnake, Milksnake, Common Watersnake, Dekay's Brownsnake, Plains Gartersnake, Common Gartersnake, and Lined Snake.

North American Racer
Coluber constrictor

DESCRIPTION

A slender body built for speed, prominent eyes, and an alert attitude are key characteristics of the North American Racer. The dorsal color of adults is solid slate blue or light brown. The throat and neck are usually bright yellow. The yellow throat grades into a light-gray or smoky-white abdomen. The Racer has a divided anal plate and smooth body scales. Adult males have longer tails than females; up to 27 percent of their total length may be tail. Racer hatchlings and juveniles differ markedly in appearance from the adults until their third summer. Young snakes have reddish-brown or black dorsal blotches on a gray background, and their white ventral surface is covered with numerous small, reddish-brown spots.

Total length of adults ranges from 90 to 152 centimeters (36 to 60 inches; Conant and Collins 1998). According to Breckenridge (1944), the largest specimen recorded in Minnesota was found in Houston County with a length of 140 centimeters (55 inches).

North American Racer, adult.
Photograph by Barney Oldfield.

No other species of snake in Minnesota is likely to be confused with an adult North American Racer with the possible exception of the Smooth Greensnake. The brilliant color of the Smooth Greensnake fades after death to a bluish gray, but a Racer the size of an adult Smooth Greensnake would still have a blotched juvenile pattern. Juvenile North American Racers can be confused with the young of several species, but the only other Minnesota species with smooth scales and a blotched pattern is the Milksnake. Milksnakes have a single anal plate in contrast to the divided plate of the Racer.

DISTRIBUTION

The North American Racer has a broad distribution in the United States. It is found from southern Maine to Florida and west to the Pacific coast. However, large gaps occur in its range in the Southwest and the upper Midwest, including northern Minnesota and the eastern Dakotas. In Minnesota, North American Racers are found as far north as Pine County and as far west as Blue Earth County, but its distribution is primarily restricted to counties along the lower Minnesota River and in the driftless area of southeast Minnesota.

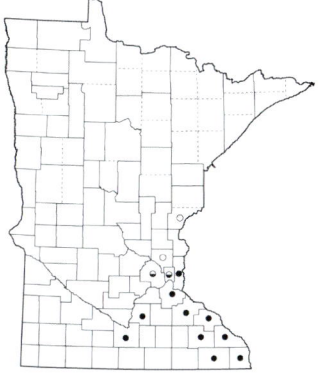

HABITAT

The North American Racer lives in a variety of open dry habitats, such as brushy areas along the edges of deciduous woodlands, grass prairies, bluff prairies, and old fields. Because these snakes primarily hunt by sight, areas of dense vegetation are not suitable. Racers hibernate in mammal burrows, rock crevices, gravel banks, stone walls, and abandoned wells. They may share these winter homes with other Racers, Timber Rattlesnakes, Western Ratsnakes, Gophersnakes, or Common Gartersnakes.

LIFE HISTORY

North American Racers become active on warm days during the last half of April, and they bask

Belly coloration of North American Racer. Photograph by Tom Jessen.

for several days near their overwintering site before leaving to forage for food. They are active, fast moving, and diurnal, and they are frequently found basking in bushes and shrubs. Racers tolerate a wide range of air temperatures, and they are often out when other snake species are under cover. According to Fitch (1963a), they may be active at temperatures of 15.5° to 32°C (60° to 90°F). They have large home ranges of 9.7 to 10.5 hectares (24 to 26 acres; Fitch 1963a).

North American Racers are opportunistic feeders consuming a large variety of food items. They eat small mammals, birds, reptiles, amphibians, and large insects. Prey items also include snails, spiders, and bird eggs. LeClere (2013) reported finding fecal masses that were entirely made up of locust exoskeletons. In Minnesota, Six-lined Racerunners are an important food source for North American Racers where available.

While hunting, the North American Racer holds its head 15 to 20 centimeters (6 to 8 inches) above the ground to gain elevation and to get a better view of its surroundings. The snake makes a quick dash to catch and seize prey when it is spotted. The Racer is not a constrictor as the scientific name suggests; however, it uses its body to press struggling prey to the ground. The snake repeatedly bites and chews on the prey item until it is sufficiently subdued and then swallows it whole.

Details on North American Racer reproduction in Minnesota are scant. Field studies by Fitch (1963a) in Kansas provide considerable information. Mating generally takes place during May and early June. The male trails the female by scent, and he initiates courting by moving alongside her with a jerking

North American Racer, hatchling. Photograph by Barney Oldfield.

North American Racer, juvenile. Photograph by Minnesota Department of Natural Resources—Carol D. Hall.

motion. After the female becomes passive, the male places his tail under hers to achieve copulation. In late June or early July the female lays a clutch of 8 to 21 eggs under logs, in rotting stumps, or inside mammal burrows. One clutch per year is normal. The cream-colored eggs are elliptical in shape with a leathery shell, and they average 3.0 by 2.0 centimeters (1³⁄₁₆ by ¾ inches). Eggs hatch in 43 to 65 days depending on ambient temperatures. Hatchling Racers emerge from the nest in late August or early September at a total length of 20 to 35 centimeters (8 to 14 inches). They reach reproductive maturity within two to three years (Fitch 1963a).

North American Racers primarily rely on flight to escape predators, but they will defend themselves. When surprised, they make a dash for cover at high speed; they have been clocked at 6.5 kilometers per hour (4 miles per hour; Vogt 1981). If cornered, Racers rapidly vibrate their tail and readily strike at the adversary.

Natural enemies include hawks, American crows, red foxes, raccoons, and striped skunks, but human activities take the largest toll. Vehicles kill countless individuals each year, and losses result from habitat destruction.

REMARKS

There are eleven recognized subspecies of North American Racer across the United States. The various subspecies are primarily differentiated by body color including blue, brown, black, gray, tan, and mottled. The Blue Racer (*Coluber constrictor foxi*) is the subspecies found in Minnesota (Crother 2012). The North American Racer is classified as special concern and is listed as a Species of Greatest Conservation Need by the Minnesota DNR (2006, 2013).

Ring-necked Snake

Diadophis punctatus

DESCRIPTION

The Ring-necked Snake is a small, shiny blue-black or gray-black snake with a conspicuous yellow or orange ring around its neck. The light yellow or bright orange belly contrasts markedly with its dark back. The orange may grade into brick red under the tail. The bright abdominal color is accented with many small black spots, although they may be absent in individuals from northern Minnesota. Ring-necked Snakes have smooth dorsal scales and a divided anal plate.

The total length of the Ring-necked Snake is 25.4 to 38 centimeters (10 to 15 inches) (Conant and Collins 1998). Adult females are generally larger than males.

Red-bellied Snakes and Dekay's Brownsnakes are similar in size and dorsal coloration, but they have keeled body scales, lack a brightly colored necklace, and have unspotted bellies.

Ring-necked Snake, adult. Photograph by Allen Blake Sheldon.

DISTRIBUTION

The Ring-necked Snake has an extensive continental range. It is found from Maine to Minnesota and south to Florida, Texas, and Arizona. Ring-necked Snakes are also found in the Pacific coastal states. In Minnesota, Ring-necked Snakes are encountered in scattered populations along the eastern edge of the state.

HABITAT

In southeastern Minnesota, Ring-necked Snakes occupy south- or west-facing hillsides. They can be found under rocks in forested areas or on steep bluff prairies. Distribution within apparently suitable habitat is spotty. Populations may be found on one hill but be absent on a nearby hill. In northern Minnesota, they are occasionally found under rocks, logs, or bark in damp deciduous forests. Abundant ground cover is a critical requirement of this species. Rock crevices and mammal burrows below frost line are used for overwintering by Ring-necked Snakes. During the hot part of summer they estivate in similar locations.

LIFE HISTORY

In Minnesota, the activity season for Ring-necked Snakes begins in April and ends in October. Ring-necked Snakes are extremely secretive and rarely bask. They depend on warm soil under shallow rocks to thermoregulate.

Belly of Prairie Ring-necked Snake. Photograph by Minnesota Department of Natural Resources—Carol D. Hall.

Under the cover of darkness they forage for earthworms, slugs, sow bugs, grubs, and spiders. Small salamanders, frogs, lizards, and snakes are used as food where available (Ernst and Barbour 1989). Ring-necked Snakes rely heavily on their sense of smell to locate food, and they are powerful constrictors once the prey animal is seized.

Ring-necked Snakes are gregarious in the spring and fall, when numerous individuals can be located under a single rock. Courtship and mating occur during these aggregations. Females lay eggs during early summer in moist sand under bark or rocks. Rotting logs and decaying tree stumps are also used as nesting sites. Clutch size varies from 2 to 10 eggs, and the average is 4. Eggs average 1.3 by 2.8 centimeters (½ by 1 inch) and have thin, leathery shells with a bumpy texture. Forty to 60 days of development within the egg are required; however, Peterson (1956) reported a female that gave birth to six live young. Communal nesting has been reported, and as many as 55 eggs have been found in one nest (Blanchard 1936). Hatchlings are 8.5 to 11.5 centimeters (3⅜ to 4½ inches) long and become sexually mature in two or three years.

Individuals rarely attempt to bite when handled, but they do writhe and discharge a foul-smelling musk. When threatened by a predator, they corkscrew their tail, exposing the bright-colored underside, while they keep their head hidden under body coils. Ring-necked Snakes serve as food for larger snakes, such as North American Racers and Milksnakes, as well as for hawks, owls, and small mammals.

Ring-necked Snake, hatchling. Photograph by Allen Blake Sheldon.

Ring-necked Snake habitat in southeast Minnesota. Photograph by Allen Blake Sheldon.

REMARKS

Two subspecies of Ring-necked Snakes occur in Minnesota. The Prairie Ring-necked Snake *(Diadophis punctatus arnyi)* is found in the southeastern counties. The belly is brightly colored with numerous black spots. The Northern Ring-necked Snake *(D. p. edwardsi)* is found in Pine County and north. The northern form has a paler abdomen with meager or no spotting. This inoffensive little snake has been nicknamed the "corkscrew snake" due to its habit of tail coiling when alarmed.

Plains Hog-nosed Snake
Heterodon nasicus

DESCRIPTION

The Plains Hog-nosed Snake, also nicknamed "prairie rooter," is a medium-sized, stout-bodied snake with a shovel-shaped snout used for digging in loose, sandy soil. The flattened, upturned rostrum bears a sharp point and a dorsal ridge. The ground color of this species is tan or buffy gray. A single row of 35 to 40 large, dark-brown blotches runs along the back. Two rows of smaller spots are found on each side of the body. The blotches form a fractured ring pattern on the tail. Two oblong, dark-brown blotches are located on the back of the neck. A dark bar connects the eyes and extends from each eye to the corner of the mouth on either side of the head. The ventral surface of the snake bears a large, wide, black or blue-black stripe that is edged in cream or yellow, and the underside of the tail is black. The dark belly coloration of freshly shed individuals often shows an iridescent sheen. Body scales are keeled, and the divided anal plate is generally

Plains Hog-nosed Snake, adult. Photograph by Minnesota Department of Natural Resources—Erica Hoagland.

a yellowish hue. Proportionally, males have longer tails than females. Ground coloration is lighter in juveniles, which causes the pattern to be more striking than in adults; otherwise, they are very similar in appearance.

Adult length of this species generally falls between 38 and 63.5 centimeters (15 and 25 inches) (Conant and Collins 1998). The largest reported individual in Minnesota was a female with a total length of 99 centimeters (39 inches) (Oldfield and Moriarty 1994).

The Eastern Hog-nosed Snake, also found in Minnesota, has a less upturned snout, and the underside of its tail is lighter in color than the adjacent belly.

DISTRIBUTION

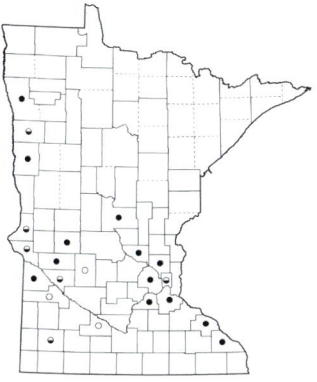

The range of the Plains Hog-nosed Snake forms a broad band running north and south across the central United States. Commencing in southwest Manitoba and southeast Alberta, the range extends south across the Great Plains into northern Mexico. Disjunct populations are located in Illinois, Missouri, and eastern Texas. In Minnesota, the Plains Hog-nosed Snake has a scattered distribution but generally is found along the western edge of the state and across the central section. Several counties in the southeast and southwest have isolated populations.

HABITAT

This snake is a prairie animal and prefers open, sandy or gravelly land. Well-drained loose loam or sand is needed for the snake's burrowing activities (Platt 1969). River floodplains and sand dunes are used in isolated locales in the southeastern corner of the state. Smith (1961) reported that Plains Hog-nosed Snakes hibernate below the frost line in mammal burrows, but specific information concerning Minnesota wintering sites is unavailable.

LIFE HISTORY

The activity period for this snake extends from early May through late September, and short intervals of estivation may occur during extreme summer conditions (Ernst and Barbour 1989). The Plains

Plains Hog-nosed Snake playing dead. Photograph by Christopher E. Smith.

Hog-nosed Snake is primarily diurnal, concentrating its activity to morning hours and late afternoon, while nights are often spent in self-excavated burrows. Recent observations (Hoaglund and Smith 2012) have found Plains Hog-nose Snakes active on warm nights. The snake digs in loose soil by thrusting its shovel-like nose side to side while pushing forward (Platt 1969).

Plains Hog-nosed Snakes consume toads, frogs, salamanders, lizards, small snakes, mice, and shrews. They also eat a variety of eggs, including those of turtles, lizards, snakes, and ground-nesting birds (Iverson 1990; Platt 1969). They actively excavate Blanding's Turtle nests and will feed on the eggs for several days (Hoaglund and Smith 2012). In Minnesota, Prairie Skinks are an important food source (Breckenridge 1944). The snake locates prey by sight and odor and then rapidly crawls up and seizes it. Struggling prey is held down with a body loop. Enlarged teeth in the back of the upper jaw help to hold and immobilize prey while toxic saliva is released and chewed into the animal.

Plains Hog-nosed Snake reproduction has not been reported on in Minnesota; however, Platt (1969) studied populations in Kansas and discovered that females mature at 20 to 22

Close-up of Plains Hog-nosed Snake. Photograph by Tony Gamble.

Hatching Plains Hog-nosed Snakes. Photograph by James E. Gerholdt.

months of age with a snout–vent length of 35 to 40 centimeters (14 to 16 inches). Most males mature at one year of age, and they wander widely searching for odor trails left by females. Mating takes place primarily in the spring, but fall encounters are likely. After a gestation period of approximately 30 days, nest excavation and egg laying take place in July. Eggs are laid singly in a row or in a cluster about 10 centimeters (4 inches) deep in a depression in damp soil. Commonly 8 to 12 eggs form a clutch, the range being 2 to 24. The white or cream eggs are elliptical with smooth, leathery shells, and their average size is 2 by 3.5 centimeters (¾ by 1⅜ inches). Development takes 47 to 75 days; young hatch from early August through September. It may take a

Plains Hog-nosed Snake digging in sand. Photograph by Minnesota Department of Natural Resources—Erica Hoagland.

hatchling 40 to 60 hours to leave the egg after first slitting it open (Munro 1949). Hatchling snakes are 14 to 19 centimeters (5½ to 7½ inches) in total length. In captivity young can be highly cannibalistic, occasionally consuming littermates soon after leaving the egg (Oldfield and Moriarty 1994).

When encountered in the wild, this species attempts to escape by crawling into a hole or burrowing into loose soil. If the snake is cornered, it hides its head under body loops. With further provocation it takes a defensive posture, spreads its neck, hisses, and repeatedly strikes with a closed mouth. If this fails to deter the intruder, the snake contorts about, vomits up any recent meal, and flips onto its back with mouth agape and tongue extended. Feces and blood may ooze from the vent. If the snake is turned over, it immediately rolls to its back again. When danger moves on, the snake rights itself from feigning death and crawls off. After a short time in captivity, they generally refuse to perform this behavior.

Predators of the Plains Hog-nosed Snake include hawks, American crows, red foxes, coyotes, and raccoons. The greatest threat to this species is the continual loss of habitat due to human activities. Also, this species is a desirable pet and is occasionally collected for that purpose, which can have a negative impact on Minnesota populations.

REMARKS

On rare occasions, the bite of the Plains Hog-nosed Snake has resulted in painful swelling and discoloration at the site of the bite (Bragg 1960; Hornfeldt and Keyler 1987). Since the saliva is mildly toxic and the snake must chew at some length to introduce the saliva, the snake poses no significant danger to humans.

Being a species dependent on prairie, the Plains Hog-nosed Snake has lost considerable habitat to agriculture and development. The greater part of the remaining populations are isolated relicts and are vulnerable to further development. The Minnesota DNR (2006, 2013) classifies this snake as special concern and as a Species of Greatest Conservation Need.

The previous common name for this species was Western Hog-nosed Snake *(H. n. nasicus)*. This subspecies of *H. nasicus* has been elevated to full species, so the common name followed that change (Crother 2012).

Eastern Hog-nosed Snake
Heterodon platirhinos

DESCRIPTION

The Eastern Hog-nosed Snake is a medium-sized, stout-bodied reptile with a sharply pointed, slightly upturned nose. The rostral plate is flattened underneath and keeled on top. Ground coloration may be gray, yellow brown, or olive brown. Two large, dark-brown spots adorn the back of the head and resemble "eyespots" when the snake flattens its neck. The rest of the body may be decorated with 20 to 30 large, dark-brown blotches running down the middle of the back. Lateral to these on either side is a row of small, black or dark-brown spots. The blotched pattern becomes a ringed design on the tail. Many individuals in Minnesota, especially older adults, lack or have indistinct blotches with exception of the eyespots. The belly is a mottled yellowish brown, dark brown, or gray. The underside of the tail is distinctly lighter colored than the adjacent abdomen. All body scales are keeled, and the anal plate is divided. Young of this species have a distinct blotched dorsal pattern. The ventral surface is black, while the underside of the neck and tail are yellow or white.

Eastern Hog-nosed Snake, adult.
Photograph by Jeffrey B. LeClere.

The average length of an adult Eastern Hog-nosed Snake ranges from 51 to 84 centimeters (20 to 33 inches; Conant and Collins 1998). The largest known Minnesota specimen was a female from Pine County with a total length of 109 centimeters (43 inches).

The snout of the Eastern Hog-nosed Snake is less upturned than that of the Plains Hog-nosed Snake. Another important distinction is that the underside of the tail of the Eastern Hog-nosed Snake is distinctly lighter, whereas in the Plains species it does not markedly differ in color or pattern from its belly. Gophersnakes have a pointed nose, but they lack the upturned rostrum found in Eastern Hog-nosed Snakes. Eastern Hog-nosed Snakes also have a single anal plate.

DISTRIBUTION

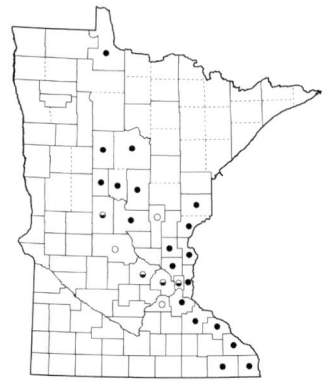

The Eastern Hog-nosed Snake ranges across the eastern United States from the southern extreme of New Hampshire south to Florida and west to Texas. This snake reaches the northwestern limit of its distribution in eastern Minnesota. Records of Eastern Hog-nosed Snakes in Minnesota come from central, east-central, and southeastern counties. There is an isolated record from Lake of the Woods County in northern Minnesota.

HABITAT

The habitat of this species consists of river floodplains, open woodlands, forest edges, and grasslands. An important component of the habitat is sandy or loamy soil. Sand beaches and adjacent wooded areas along the Mississippi and St. Croix Rivers and their tributaries provide these conditions in Minnesota. Individuals hibernate in mammal tunnels or self-dug burrows below the frost line (Platt 1969).

LIFE HISTORY

Eastern Hog-nosed Snakes emerge from hibernation early in the spring and become active by the last half of April. They actively search for food or mates aboveground during morning hours, but much of their time is spent underground. Little is known concerning their home range, but Platt (1969) reported move-

Eastern Hog-nosed Snake in gray phase. Photograph by Barney Oldfield.

ments of 858 meters (938 yards) by males during the breeding season. Reports of Eastern Hog-nosed Snakes swimming are not uncommon.

Toads are the primary prey of the Eastern Hog-nosed Snake, and this species has several adaptations for feeding on these amphibians. The modified rostrum enables the snake to search and burrow for hidden toads. Once seized, a struggling toad inflates its body and secretes toxic alkaloids from skin glands. The snake is equipped with elongated teeth in the back of its movable upper jaw that can secure a grip and deflate the toad. The snake's saliva possesses a mild toxin that helps subdue the prey (McKinistry 1978). In addition, this species is apparently immune to skin secretions of toads (Huheey 1958). Alternate food items include

Eastern Hog-nosed Snake, juvenile. Photograph by Minnesota Department of Natural Resources—Carol D. Hall.

Close-up of head of Eastern Hog-nosed Snake. Photograph by Minnesota Department of Natural Resources—Carol D. Hall.

salamanders, various species of frogs, skinks, small mammals, birds, and arthropods (Wright and Wright 1957).

Sexual maturity occurs at 18 to 21 months of age, and adults are believed to reproduce annually (Platt 1969). Males must actively search for females since they overwinter individually. Courtship and copulation occur from mid-April through May. Cream-colored, elliptical eggs with thin parchment shells are laid in loose, sandy soil or humus in protected areas from late May to early July. Average clutch size is 15 to 25 eggs; however, one Pine County female contained 61 developed eggs (Breckenridge 1944). The size of an egg ranges from 2.1 to 3.9 centimeters (¾ to 1½ inches) by 1.3 to 2.8 centimeters (½ to 1⅛ inches), and egg development takes 50 to 65 days depending on weather. Ernst and Barbour (1989) report that young emerge from the egg at a length of 16.8 to 25 centimeters (6⅝ to 9¾ inches).

The Eastern Hog-nosed Snake readily puts on a remarkable show of bluff when threatened. The snake coils its body, spreads its neck, and raises its head like a cobra. It hisses loudly and repeatedly strikes with a closed mouth. If agitated further, it goes into a death-feigning act by contorting its body, gaping its mouth, excreting feces, and rolling onto its back. However, there is a basic flaw in the performance. If the snake is turned right side up, it immediately flips on its back again. Once danger passes, the snake rolls over and crawls away. Hatchlings will perform the death-feigning act right out of the egg (Ernst and Barbour 1989).

Predators include birds of prey, small carnivorous mammals, and other species of snakes (Ernst and Barbour 1989). Uninformed observers often kill hog-nosed snakes because of their defensive display. Roadkills and habitat destruction take significant numbers of these interesting serpents.

REMARKS

The Eastern Hog-nosed Snake has a long list of nicknames, including "blow snake," "hissing adder," "puff adder," "spread-head," and "sand adder." Generally this species is demanding to maintain in captivity due to the difficulty of obtaining a constant supply of toads. The Eastern Hog-nosed Snake is listed as a Species of Greatest Conservation Need by the Minnesota DNR (2006).

Eastern Hog-nosed Snake defensive hooding. Photograph by Allen Blake Sheldon.

Milksnake

Lampropeltis triangulum

DESCRIPTION

The Milksnake is a medium-sized snake with a row of large, reddish-brown to grayish-brown saddle blotches down its back. These saddles are bordered in black and alternate with one or two rows of smaller spots along the sides of the body. The dorsal ground color is light gray or light brown, and a light Y- or V-shaped pattern is generally present on the back of the neck. The top of the head is diversely marked, while the chin is a uniform dusty white. The light-colored belly is boldly marked in a checkerboard pattern with an irregular placement of numerous solid black squares. Body scales are smooth, and the anal plate is entire. Hatchling and juvenile snakes are patterned like adults, but the dorsal blotches are often bright red and contrast sharply with a light-gray background.

The majority of adult Milksnakes are 61 to 90 centimeters (24 to 36 inches) in length (Conant and Collins 1998). The largest known Minnesota individual, from Goodhue County, was measured at 107 centimeters (42 inches).

Milksnake, adult. Photograph by Allen Blake Sheldon.

Minnesota snakes that might be confused with Milksnakes include Western Foxsnakes, Northern Watersnakes, young North American Racers, and young Western Ratsnakes. All of these species, however, have divided anal plates. Young Milksnakes tend to have brighter colors than the other species.

DISTRIBUTION

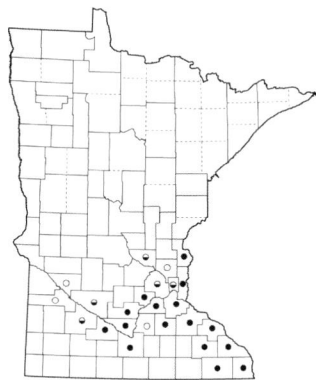

The range of the Milksnake, which includes several subspecies, is one of the largest of any North American snake. It is found from southern Maine south to Florida and west to the Rocky Mountains, virtually spanning the eastern three-fourths of the United States. In Minnesota, the Milksnake is found in watershed regions of the St. Croix, Mississippi, and Minnesota Rivers in the southern third of the state.

HABITAT

Minnesota Milksnakes are primarily found in deciduous wooded valleys and hills often in association with rocky areas near water sources. Woodlots, abandoned rock quarries, and old farmsteads where rodents are common are preferred habitat. Congregations of Milksnakes are found in rocky upland sites in spring and fall, and they often move to lower ground during the summer to hunt. This snake also appears with considerable regularity within towns, where it is found in basements and around foundations of old homes and buildings. Hibernation sites are in rock outcrops, mammal burrows, cisterns, and foundations of old buildings.

LIFE HISTORY

Milksnakes emerge from hibernacula by mid-April. Individuals can be found basking near these sites in early May and again in September (Vogt 1981). Seven adult Milksnakes were found massed together under one rock in a quarry one May evening (Oldfield and Moriarty 1994). Because they are very secretive, the easiest way to find Milksnakes during late spring and summer is by searching under rocks, logs, boards, and tin. On occasion they may be found crossing blacktop roads on warm summer nights. By October they return to protected locations

Milksnake, juvenile. Photograph by Allen Blake Sheldon.

to hibernate. A juvenile was observed crawling on a sidewalk in Great River Bluffs State Park (formerly O. L. Kipp State Park) on November 20 at an air temperature of 4.4°C (40°F; Oldfield and Moriarty 1994).

Milksnakes are powerful constrictors. They seize prey in their mouth and swiftly wrap body coils around it. Normally, the victim dies of suffocation before being swallowed. About 80 percent of their food volume consists of small mammals, including mice, voles, shrews, and young rats, but Milksnakes also eat small birds, bird eggs, lizards, lizard eggs, and other snakes (Ernst and Barbour 1989). Williams (1978) reported that small frogs, small fish, earthworms, slugs, and insects are also consumed. According to Vogt (1981), the primary food of hatchling Milksnakes is the young of other snakes, including Ring-necked Snakes, Common Gartersnakes, Smooth Greensnakes, Brownsnakes, and Red-bellied Snakes.

Courtship and mating take place anytime from emergence through early June. Gestation is 30 to 40 days. During the latter half of June, the gravid female selects a site in decaying humus, in rotting wood, under rocks, or in mammal burrows to lay 3 to 24 eggs. The average clutch size is 8 to 12, and the white, cylindrical eggs are 1.1 to 1.5 centimeters by 2.1 to 3.5 centimeters (⅜ to ½ inch by ⅞ to 1½ inches). By late August or early September the young snakes leave the egg to begin life on their own. Total length of newly hatched Milksnakes varies from 15 to 25 centimeters (6 to 10 inches).

When surprised, a Milksnake coils and rapidly vibrates its tail. Some individuals strike and bite fiercely, while others hide their head under body coils. Hawks, owls, and small carnivorous mammals are predators of the Milksnake. Vehicles and habitat destruction are responsible for the loss of countless numbers each year.

REMARKS

Lampropeltis triangulum triangulum (Eastern Milksnake) is the only subspecies recognized in Minnesota (Conant and Collins 1998); however, occasional individuals are found in southern Minnesota that demonstrate characteristics of the subspecies *L. t. syspila* (Red Milksnake), which is found in northern Iowa and farther south. The Red Milksnake has large, bright-red blotches and a noticeable reduction in lateral spots when compared to the Eastern Milksnake. The area of intergradation between the two subspecies is broad and probably extends into southeastern Minnesota.

The origin of the name *milksnake* comes from an erroneous belief by some farmers that a cow with poor milk production had been recently milked by this serpent. There is no basis in fact for this folklore. Rodents are plentiful around livestock feeds, and this is the reason Milksnakes are found in and near barns. Milksnakes are listed as a Species of Greatest Conservation Need by the Minnesota DNR (2006).

Common Watersnake

Nerodia sipedon

DESCRIPTION

Minnesota's only watersnake is the Common Watersnake. This heavy-bodied aquatic snake possesses a distinct head with a blunt snout. The ground color is reddish brown, tan, or gray. The neck and forepart of the body have dark-brown, dark-gray, or black bands that are wider than the spaces between them. From midbody toward the tail, these bands break into large, squarish dorsal blotches and smaller spots along the sides. The bands and blotches may be outlined with black. Irregular dark markings are found on top of the head, and the chin is off-white. Adult Common Watersnakes with dry skin are uniform dusky brown or tan with little discernible pattern. When the skin is wet, however, the pattern becomes conspicuous. The cream to yellowish belly has numerous irregularly spaced half-moons with dark borders and reddish-brown centers. On occasion, a solid gray or tan individual is found with no dorsal markings and a pale

Common Watersnake, adult. Photograph by Allen Blake Sheldon.

yellowish belly with sparse, small black specks. Body scales are strongly keeled, and the anal plate is divided. Juveniles are similarly patterned as adults, but the bands and blotches are more conspicuous due to a lighter ground color.

The adult Common Watersnake ranges in total length from 61 to 107 centimeters (24 to 42 inches) (Conant and Collins 1998). Females are larger than males (Gibbons and Dorcas 2004).

Two species of snakes that might be confused with the Common Watersnake because they are frequently found near water are the Western Foxsnake and the Massasauga. Both of these species have a row of middorsal blotches. In addition, the Western Foxsnake has a solid-colored head, and the Massasauga has a rattle on the end of its tail. The only other species in Minnesota that has bands across its body is the Timber Rattlesnake, but it has a black tail with a tan rattle.

DISTRIBUTION

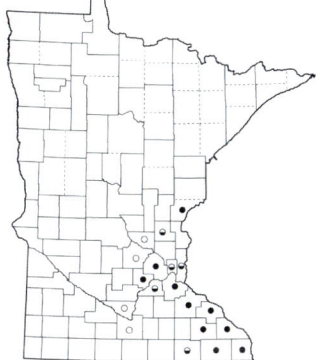

Common Watersnakes range from southern Maine south to the panhandle of Florida and west to Oklahoma and eastern Colorado. A large section of the southeastern coastal plain is devoid of this species. In Minnesota, this snake is found in a band of counties from the extreme southeast northward into Pine County. The distribution of this species follows the Mississippi, Minnesota, and St. Croix Rivers and their tributaries.

HABITAT

As the common name of this species suggests, it is found in or near water. Almost any body of freshwater, including ponds, lakes, rivers, sloughs, creeks, marshes, and bogs, provides suitable habitat (Vogt 1981). Sufficient quantities of food, cover, and basking sites in and around water are fundamental requirements. Breckenridge (1944) reported that this species hibernates in upland rock crevices and holes away from water. It may use crayfish burrows, muskrat and beaver lodges, earthen dams, and levees (Ernst and Barbour 1989).

Belly pattern of Common Watersnake. Photograph by Barney Oldfield.

LIFE HISTORY

By late April, Common Watersnakes emerge from their over-wintering sites and remain active until October. They are active both day and night, but they are primarily diurnal in spring and fall. This species basks on rocks, tree roots, beaver lodges, and ground vegetation near the water's edge, and as many as five or six individuals may pile together. At the slightest hint of danger, Common Watersnakes promptly slide into the water and swim off with only their head above the surface. They have been known to remain totally submerged for over an hour with no ill effects (Ferguson and Thornton 1984).

Common Watersnake, juvenile. Photograph by Tony Gamble.

Common Watersnakes prefer to eat ectothermic vertebrates, especially fish and amphibians. According to Johnson (2000), up to 95 percent of their diet consists of small fish. Frogs, tadpoles, toads, and salamanders are a significant food source when available. Gibbons and Dorcas (2004) report that over 80 different species of fish and 30 species of amphibians have been found in the Common Watersnake's diet. Small mammals, crayfish, snails, slugs, insects, spiders, leeches, earthworms, and other snakes are also taken (Ernst and Barbour 1989).

Common Watersnake swimming.
Photograph by Jeffrey B. LeClere.

Common Watersnakes are active hunters; they swiftly pursue prey, which they seize in their mouth and swallow whole.

This species reaches sexual maturity at two to three years of age. Courtship and mating take place in spring after emergence. Several males may simultaneously court a single female, either in water or on land. The male moves alongside and on top of the female, periodically pressing his chin against her back and neck. Once copulation is achieved, they may remain together for several hours (Ernst and Barbour 1989). A single annual litter is produced after a gestation period of approximately 60 days. The developing embryos in the female's uterus are nourished through a placenta (Conway and Fleming 1960). The female gives live birth to an average of 25 (average range of 10 to 48) young during August and early September. Extraordinary large litter sizes have been recorded, 99 being the greatest (Slevin 1951). Newborn snakes range in length from 19 to 25 centimeters (7½ to 10 inches).

Common Watersnakes have a particularly bad disposition. If cornered by an intruder, they flatten their head, vibrate their tail, repeatedly strike, and bite viciously by holding on and chewing. Although the bite is not dangerous, the wound may bleed excessively due to an anticoagulant component of their saliva. Along with all of these actions, they spray a foul-smelling fluid from their cloaca.

The Common Watersnake has many predators, including Snapping Turtles, American Bullfrogs, hawks, American crows, gulls, herons, bitterns, great egrets, vultures, raccoons, mink, and large predatory fish. Juvenile snakes are especially vulnerable as prey.

REMARKS

The common name was Northern Watersnake but was changed to Common Watersnake in Crother (2012) so that a species and subspecies did not have the same common name. Four subspecies of the Common Watersnake are recognized by herpetologists. *Nerodia sipedon sipedon* (Northern Watersnake) is the subspecies found in Minnesota (Conant and Collins 1998).

Humans frequently kill Common Watersnakes on sight, believing they are dangerous "water moccasins." The venomous Cottonmouth (commonly called "water moccasin") reaches its northern distributional limit in southern Missouri, several hundred miles from the Minnesota border. There is one record of a Cottonmouth from Winona. The snake was found on a barge that had come upriver (Cochran 2008). Common Watersnakes are often persecuted by fishermen who believe they consume too many game fish. In actuality, this species reduces competition between fish by feeding on small, diseased, and injured fish. The snake itself provides food for larger bass and pike. They serve as an important link in the natural balance of aquatic ecosystems.

Smooth Greensnake

Opheodrys vernalis

DESCRIPTION

The Smooth Greensnake is a small, beautiful, emerald-green colubrid with a slender body and a pristine white or pale-yellow abdomen. Infrequently, individuals are light brown or tan instead of green. Their bright-red tongue has a black tip. Body scales are smooth, and the anal plate is divided. Hatchlings are olive green. Smooth Greensnakes reach a length of 36 to 51 centimeters (14 to 20 inches) (Vogt 1981).

Shortly after death, the Smooth Greensnake's coloration fades to pale blue, superficially resembling the North American Racer and confusing identification. Racers the size of adult Smooth Greensnakes still bear a juvenile blotched pattern.

DISTRIBUTION

Smooth Greensnakes are found from southeastern Canada and Maine south to Virginia and west to the Rocky Mountains. The range of this species is broken into numerous isolated

Smooth Greensnake, adult.
Photograph by Jeffrey B. LeClere.

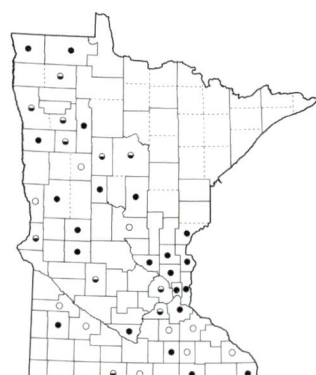

populations across the Great Plains. Excluding the Arrowhead region, records of Smooth Greensnakes are widely scattered across Minnesota.

HABITAT

These snakes are found in prairies and meadows, along the edges of mixed hardwood and pine forests, near the edges of marshes and bogs, and on dry hillsides. They hibernate below the frost line at soil depths greater than 15 centimeters (6 inches). Lang (1969) discovered large numbers of Smooth Greensnakes using abandoned ant mounds as communal hibernacula with other species, including Red-bellied Snakes, Common Gartersnakes, and Plains Gartersnakes.

Color morphs of Smooth Greensnake. Photograph by Minnesota Department of Natural Resources—Carol D. Hall.

LIFE HISTORY

The active season for this species in Minnesota begins in mid-April and extends through September. They are primarily diurnal snakes but are occasionally found crossing roads on warm, rainy evenings (Vogt 1981). Smooth Greensnakes climb into shrubs for foraging or basking, but they also spend considerable time on the ground crawling in vegetation and hiding under ground cover (T. R. Johnson 2000).

Smooth Greensnakes consume invertebrates, including spiders, slugs, centipedes, millipedes, crickets, grasshoppers, moth and butterfly larvae, and beetles (Ernst and Barbour 1989). They seize prey with their mouth and swallow it whole. This snake is especially vulnerable to pesticides due to its arthropod diet. Minton (1972) collected two specimens from a sprayed field that died within two weeks of collection.

Mating probably occurs during May, although it has also been observed in August (Dymond and Fry 1932). Smooth Greensnakes nest during late July and August; 3 to 11 eggs is the normal clutch size. Communal nesting is likely, as large numbers of eggs (up to 31) have been found at one site (F. C. Cook 1964). Nests are located in mounds of rotting vegetation, decomposing logs, tree hollows, and sawdust piles. Eggs are cylindrically shaped with thin, white shells and measure 2.3 by 1.3 centimeters (⅞ by ½ inch). Hatching occurs about a month after egg deposition, and the young snakes are 10 to 15 centimeters

Smooth Greensnake, juvenile.
Photograph by Allen Blake
Sheldon.

(4 to 6 inches) in length when they leave the eggs. Egg development periods as short as four days have been reported, indicating that the female has the ability to retain maturing eggs within her body (Vogt 1981). Hatchlings retain residual egg yolk, which allows them to go without feeding for the first two weeks. Young snakes begin foraging for food after they shed their skin the first time.

The Smooth Greensnake's main line of defense is cryptic coloration, which makes it very difficult to detect in green vegetation. When threatened, it raises its head, gapes, and strikes, but its small mouth is only capable of inflicting a minor scratch. Predators include birds, mammals, and larger snakes. Habitat destruction and pesticides are significant threats to this species.

REMARKS

Conant and Collins (1998) do not list subspecies for this snake. Some people refer to this harmless, inoffensive little serpent as a "grass snake."

The Smooth Greensnake is listed as a Species of Greatest Conservation Need by the Minnesota DNR (2006).

Western Ratsnake

Pantherophis obsoletus

DESCRIPTION

The adult Western Ratsnake is a large, impressive constrictor with a white chin and throat and weakly keeled dorsal scales. A pattern of middorsal blotches is generally discernible. The skin of the back and sides, especially on the anterior third of the snake, shows small flecks of red, orange, or white. The belly is dark gray or brown with indistinct darker specks or checkerboard blotches. The cross-sectional shape of the Western Ratsnake's body is very similar to a loaf of bread with vertical sides, a rounded top, and a flat bottom. Western Ratsnakes have a divided anal plate. Young Western Ratsnakes are distinctly different from adults. They have a gray ground color with a bold pattern of dark blotches down the length of the body. A dark band runs from the eye to the corner of the mouth. The ventral surface of a juvenile shows a checkerboard pattern of dark brown or black on a light background. After two years of growth, juvenile coloration begins changing to that of the adult.

The average length of an adult is 107 to 183 centimeters (42 to 72 inches; Conant and Collins 1998).

Western Ratsnake, adult.
Photograph by Jeffrey B. LeClere.

Young Western Ratsnakes are easily confused with the young of several other Minnesota snakes. Juvenile North American Racers and Milksnakes also have a blotched pattern, but they have nonkeeled scales. Hatchling Gophersnakes have strongly keeled scales, a pointed nose, and a single anal plate. The most reliable way to distinguish a juvenile Western Ratsnake from a Western Foxsnake is to count ventral scales. Western Ratsnakes have 221 or more ventral scales, while Western Foxsnakes have 216 or less. Melanistic Gartersnakes have also been confused with Western Ratsnakes.

DISTRIBUTION

The Western Ratsnake has a broad distribution across the eastern United States, extending from southern New England south to the Florida Keys and west to Texas and extreme southeastern Minnesota. This species is extremely rare in Minnesota. Only four documented locations have been reported over a span of 50 years (Smith and Kozack 2011). Counties of record include Fillmore, Houston, and Winona. Recent records are all from Houston County. Ongoing research by the Minnesota Biological Survey has located two populations with over 40 individuals in Houston County (J. LeClere, pers. comm.).

HABITAT

The Western Ratsnake is a woodland species making its home in rocky, timbered uplands, wooded valleys, and forests on the backside of south-facing bluffs. According to Vogt (1981), Ratsnakes in Wisconsin are found in moist, wooded east and north slopes near bluffs along rivers. Known records indicate that the only suitable Western Ratsnake habitat in Minnesota occurs in the extreme southeastern counties where forested hills are dissected by rivers and streams. Ratsnakes hibernate in deep, rocky crevices below the frost line. They very likely share dens with North American Racers, Gophersnakes, and Timber Rattlesnakes.

LIFE HISTORY

Western Ratsnakes emerge from hibernation in late April and May, and after several days of basking in nearby shrubs and trees

Arboreal activity of Western Ratsnake. Photograph by Minnesota Department of Natural Resources—Carol D. Hall.

(J. LeClere, pers. comm.), the diurnal snakes move into adjacent woodlands, where they are often found 6 to 12 meters (6½ to 13 yards) off the ground in oak and hickory trees (Vogt 1981). They regularly take refuge in cavities of hollow trees. By early October the snakes return to hibernation dens.

Approximately two-thirds of a Western Ratsnake's diet consists of small mammals, such as mice, chipmunks, squirrels, rabbits, and shrews (Fitch 1963b). Birds, bird eggs, and nestlings are also significant food sources. Large prey is seized in the mouth and quickly suffocated by constriction. Small prey may be grabbed and swallowed alive. Juvenile Western Ratsnakes eat amphibians, lizards, and invertebrates (T. R. Johnson 2000).

Mating probably occurs in May and early June. A clutch of 6 to 30 eggs (10 to 14 average) is laid in snags, hollow logs, stumps, or sawdust piles during June or early July (T. R. Johnson 2000). The white, leathery eggs are 3.5 to 5.5 centimeters (1⅜ to 2⅛ inches) in length and 1.6 to 3.0 centimeters (⅝ to 1⅛ inches) in width. Eggs generally adhere to each other in the nest (Ernst and Barbour 1989). Hatchlings emerge after 60 to 75 days, and they are 28 to 41 centimeters (11 to 16 inches) in length.

Western Ratsnakes freeze when danger threatens, and they

Western Ratsnake, juvenile.
Photograph by Jeffrey B. LeClere.

rely on their cryptic coloration to avoid detection. If cornered, they vibrate their tail, elevate their head, and strike repeatedly. A large snake can deliver a respectable bite. Adult Western Ratsnakes have few enemies other than hawks. Young Ratsnakes are eaten by carnivorous mammals, birds of prey, and other species of snakes.

REMARKS

A common nickname for this species is the "pilot blacksnake." This misnomer is derived from unfounded folklore that claims this species of snake leads rattlesnakes away from danger or back to their dens in the fall.

It is unfortunate that so little is known of this impressive serpent in Minnesota. Suitable climatic conditions and lack of appropriate habitat apparently restrict this species within the state. The Western Ratsnake is classified as threatened and is listed as a Species of Greatest Conservation Need by the Minnesota DNR (2006, 2013).

The Western Ratsnake genus was changed from *Elaphe* to *Pantherophis* after molecular studies (Utiger et al. 2002) showed that *Elaphe* was actually comprised of eight different genera.

Western Foxsnake

Pantherophis ramspotti

DESCRIPTION

The mature Western Foxsnake is a medium to large snake with an unmarked head and a blotched dorsal pattern. The head, with a rounded nose, is solid brown or reddish brown. There are 34 to 43 large, reddish-brown, dark-brown, or black middorsal blotches, and a row of alternating smaller blotches along each side of the body. The dorsal ground color is sooty brown or tan. The belly of the Western Foxsnake is pale yellow with numerous brown or black, rectangular blotches. This stout-bodied snake has weakly keeled scales and a divided anal plate. Young Fox-snakes have a lighter ground color, and the dorsal blotches are bordered with black. They also have a dark bar across the top of the head connecting the eyes, and a bar from the eye to the corner of the mouth on either side of the head.

The average adult Western Foxsnake is 91 to 137 centimeters (36 to 54 inches) in length (Conant and Collins 1998). A large

Western Foxsnake, adult. Photograph by Allen Blake Sheldon.

individual found in Goodhue County had a total length of 155 centimeters (61 inches).

Western Foxsnakes are frequently confused with several other Minnesota species. Gophersnakes have a boldly marked, pointed head and a single anal plate. Milksnakes have a single anal plate, and their scales lack keels. Rattlesnakes have a rattle on the end of their tail and a distinctly shaped head with vertical pupils. Juvenile Western Foxsnakes are also easily confused with the young of several other Minnesota species. (See the Western Ratsnake account for a discussion of how to distinguish these juvenile snakes.)

DISTRIBUTION

The range of the Western Foxsnake is limited to the north-central United States and extreme southern Ontario. It is found from eastern Michigan across Wisconsin to eastern Nebraska, and south to Illinois and northern Missouri. In Minnesota, the range of the Western Foxsnake covers the southern half of the state; the northernmost record is from Pine County. It is broadly distributed along major river valleys, including the Mississippi, Minnesota, St. Croix, and Rock.

HABITAT

Western Foxsnakes are found in river-bottom forests, upland hardwoods, pine barrens, and prairies. They are rarely found far from rivers or streams. Rock crevices, mammal burrows, and other suitable areas below the frost line serve as overwintering sites. Vogt (1981) found 166 foxsnakes hibernating in an abandoned well in central Wisconsin, where many of them were totally submerged in the water. They will also use other man-made structures. Jessen (1991) reported an old railroad bridge supporting a large hibernacula.

LIFE HISTORY

Western Foxsnakes emerge from hibernation during the last two weeks of April. Although primarily diurnal, they are often found crossing highways on warm, rainy evenings. Foxsnakes are very capable climbers but spend considerable time on the ground. They often take cover under logs, boards, or discarded tin.

Breeding Western Foxsnakes at hibernacula. Photograph by Tom Jessen.

Western Foxsnakes hunt for mice, chipmunks, ground squirrels, and other small mammals (Vogt 1981). They also feed on birds and bird eggs, either on the ground or in trees. Being powerful constrictors, they quickly subdue larger prey items by suffocation with body coils before swallowing. Hatchling Foxsnakes feed on invertebrates, small frogs, and juvenile rodents.

Western Foxsnakes mate during late April and early May. About 30 days later the female finds a desirable nesting site under logs in damp soil, rotting humus, or woodlot sawdust piles, where she deposits a clutch of 8 to 27 eggs (Vogt 1981). The elongated, leathery eggs average 4.5 by 2.8 centimeters (1¾ by 1⅛ inches) and frequently adhere to each other (Vogt 1981). Communal nesting by several females has been reported (Ernst and Barbour 1989). The egg development period varies from 35 to 78 days depending on ambient temperatures. Hatchling Western Foxsnakes are 23 to 31 centimeters (9 to 12 inches) in total length.

When surprised by an enemy, a Western Foxsnake rapidly vibrates its tail and repeatedly strikes from a coiled position, attempting to bite the intruder. A freshly captured Foxsnake releases musk from glands at the base of its tail. This musk has an odor similar to that of a red fox, which may be the origin of the snake's common name.

Hawks capture and feed on adult snakes, and juveniles fall prey to carnivorous mammals, birds, and other snakes. The greatest loss of Western Foxsnakes is due to vehicles and habitat destruction. Moreover, the Western Foxsnake is occasionally collected for the hobby trade.

REMARKS

The Western Foxsnake has an unfortunate identification problem. It is often mistaken for a rattlesnake because it vibrates its tail when alarmed. The solid-colored reddish-brown head leads to its incorrect identification as a copperhead (a species not found in Minnesota). As a result, Western Foxsnakes are dispatched as venomous serpents by the uninformed. Additional misnomers applied to the Foxsnake are spotted adder, pinesnake, and bullsnake. Due to habitat alteration and collection pressures, the Minnesota DNR (2006) lists the Western Foxsnake as a Species of Greatest Conservation Need.

In Oldfield and Moriarty (1994), the scientific name of the Western Foxsnake was *Elaphe vulpina*. The genus was changed from *Elaphe* to *Pantherophis* after molecular studies showed that *Elaphe* was actually comprised of eight different genera (Utiger et al. 2002). The specific epithet was changed to *ramspotti* after a reevaluation of the Eastern and Western Foxsnakes by Crother et al. (2011). There is still some debate on the separation between the Eastern and Western Foxsnakes in Minnesota, which will not be settled until further DNA analysis is undertaken.

Shed Western Foxsnake skins at hibernacula. Photograph by Minnesota Department of Natural Resources—Christi Spak.

Gophersnake
Pituophis catenifer

DESCRIPTION

The Gophersnake, also known as the Bullsnake, is the longest Minnesota snake, reaching a length of more than 183 centimeters (6 feet). It is a powerful, stout-bodied constrictor with a relatively small head. The yellow head has many black or near-black markings, including a bold stripe from the eye to the corner of the mouth. There are prominent vertical lines on the upper lip, and the head is pointed with an elongated, protruding rostral plate. Ground color is straw yellow or light brown, and there are 38 to 53 black, dark-brown, or reddish-brown blotches down the midline. The blotched pattern is less obvious near the head but becomes bolder near the tail. The tail displays black or reddish-brown bands. The chin and belly of the Gophersnake are pale yellow, and the belly is stamped with numerous square or rectangular dark spots. Many individuals show a pronounced color change from head to tail. Body scales are keeled, and the anal plate is single. The pattern of juvenile Gophersnakes is similar to adults, but their overall coloration is lighter.

Gophersnake, adult. Photograph by Allen Blake Sheldon.

The average adult Gophersnake ranges in length from 95 to 183 centimeters (37 to 72 inches; Conant and Collins 1998). The largest known Minnesota Gophersnake was a female from Wabasha County that measured 188 centimeters (74 inches).

The Western Foxsnake, which is often confused with the Gophersnake, has an unmarked head with a rounded nose. Also, Western Foxsnakes have divided anal plates. Timber Rattlesnakes have an unmarked head, a banded body pattern, and a black tail with a tan rattle.

DISTRIBUTION

The Gophersnake ranges from western Wisconsin and Illinois south to Texas and west to the Pacific coast. Its range is extensive and includes the entire western United States, portions of southern Canada, and northern Mexico. In Minnesota, the Gophersnake is found in the southern half of the state, and many of the records come from counties bordering the Mississippi, Minnesota, and St. Croix Rivers. The isolated records from northwestern Minnesota have been found on the glacial beach ridge prairies of Polk and Pennington Counties.

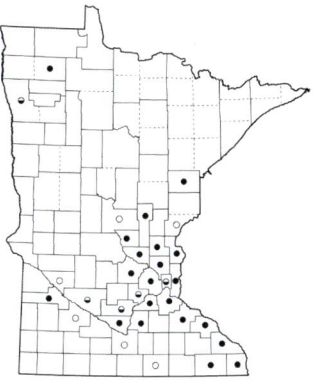

HABITAT

The Gophersnake is a snake of open country. It occupies native prairies, old fields, pastures, oak savannas, and bluff prairies that are located on steep hillsides. This species thrives in sandy-soil habitats where burrowing rodents are common. Hibernation occurs in mammal burrows, well below the frost line or deep in rock fissures in bluff areas (Moriarty and Linck 1997; Schroder 1950; Vogt 1981).

Gophersnakes hatching.
Photograph by Barney Oldfield.

LIFE HISTORY

Gophersnakes emerge from hibernation in late April or May. Hibernacula may be shared with other species, such as North American Racers, Western Ratsnakes, and

Gophersnake hibernacula entrance. Photograph by Minnesota Department of Natural Resources—Carol D. Hall.

Timber Rattlesnakes (Vogt 1981). This species is principally diurnal and can climb, but it spends much time foraging and resting in rodent burrows. Gophersnakes are fond of basking on warm sand, rock, or pavement. Their active season generally ends in late September or early October.

Gophersnakes use a large area during the summer. During a reintroduction project, Moriarty and Linck (1997) found that an individual had an average home range of 6 hectares (18 acres) but used over 200 hectares (500 acres) during the active season. Some snakes traveled over 2 kilometers (1¼ miles) during the summer.

Gophersnake digging in gopher mound. Photograph by Minnesota Department of Natural Resources—Carol D. Hall.

Gophersnake habitat.
Photograph by Tony Gamble.

Gophersnakes eat gophers and a variety of other small mammals, including mice, voles, ground squirrels, and tree squirrels. Other food items include frogs, ground-nesting birds, and bird eggs. Larger rodents are seized by the mouth and quickly suffocated by coils of the snake's body before being swallowed. Rodents encountered in burrows are pressed against the tunnel wall by the snake until they succumb. The snake itself is at risk when trying to subdue a large rodent, and may be badly bitten or, in rare instances, killed in the process (Haywood and Harris 1972). Eggs and small helpless prey are swallowed directly without constriction.

Gophersnakes mate during May. During the breeding season, a male may engage in a combat bout with another male to establish dominance (Shaw 1951). When a male finds a receptive female, he crawls alongside and on top of her in a jerking, spasmodic manner. Generally she lies still, while elevating and slowly waving her tail. He may grasp her neck, back, or head with his mouth before positioning his vent adjacent to hers to achieve copulation. Copulation may take an hour or longer (Collins 1974).

During June or early July females deposit a single clutch of 3 to 24 eggs (average of 12) in a self-excavated nest. The location of the nest may be out in the open or under a large rock or log in loose, sandy soil. Communal nests have been reported, although they are not common (Burger and Zappalorti 1986). The whitish eggs are elliptical with rough, leathery shells, and they average 4.0 by 5.0 centimeters (1½ by 2 inches). Eggs often adhere to each other, and they gain in width and mass during the development period. Guthrie (1926) reported an egg development period of 56 days, although periods as long as 100 days have been noted (Ernst and Barbour 1989). Hatchling length is 25.5 to 44 centimeters (10 to 17 inches). The skin is shed for the first time at 7 to 10 days of age.

When encountered in the wild, Gophersnakes make every attempt to escape, often diving for the first available mammal burrow. If left without a retreat, they can put on an impressive display of fierceness. An agitated Gophersnake coils, rapidly vibrates its tail, hisses loudly, and strikes repeatedly. In dry leaves the vibrating tail makes a sound similar to a rattlesnake. No other Minnesota snake can hiss as loudly. With open mouth, they forcefully expel air, causing the epiglottis, which covers the opening to the trachea, to vibrate. Their epiglottis is enlarged and may help amplify the sound (Martin and Huey 1971). Although nonvenomous, a large adult is capable of delivering a painful bite.

Adult Gophersnakes have few predators, although it is probable that a large hawk or predatory mammal may occasionally take one. Young snakes fall prey to birds, mammals, and larger snakes, such as North American Racers and Milksnakes. The loss of Gophersnakes to road traffic and habitat destruction is far more significant, and many snakes are needlessly destroyed by humans due to fear and misinformation.

REMARKS

There are six recognized subspecies of *Pituophis catenifer* (Crother 2012). The subspecies of Gophersnake found in Minnesota is the Bullsnake *(P. c. sayi)*.

The Gophersnake (Bullsnake) is an extremely beneficial predator, eliminating numerous pocket gophers and ground squirrels. Landowners battling rodent populations in croplands or grasslands benefit the rodents by killing even a single snake.

The Bullsnake name may have originated from the snake's "bullish" defensive behavior. The longevity record for a captive is just short of 22½ years (Bowler 1977). The Minnesota DNR (2006, 2013) classifies this species as special concern and lists it as a Species of Greatest Conservation Need.

Dekay's Brownsnake

Storeria dekayi

DESCRIPTION

The Dekay's Brownsnake is a small, inoffensive snake of Minnesota's woodlands. Ground color varies in earth tones from grayish brown to reddish brown, and there is usually a light middorsal stripe about four scales in width bordered by a parallel row of small, dark spots. These spots are often connected by a thin, dark line, which borders the middorsal stripe on either side, and a dark bar may connect adjacent spots across the stripe. A relatively large dark spot is located on either side of the neck, and there may be a smaller dark spot under each eye and a dark stripe behind the eye. The pale-pink or cream belly has pinpoint specks on the edges of the ventral plates. Dorsal scales are strongly keeled, and the anal plate is divided. Juvenile Brownsnakes have a light-yellow or white band around the neck. The body pattern of juveniles is often indiscernible due to a widespread dark coloration.

The Dekay's Brownsnake rarely exceeds 35.5 centimeters (14 inches) in total length, and an average adult length ranges from 23 to 33 centimeters (9 to 13 inches; Conant and Collins 1998).

Other small snakes are occasionally confused with Dekay's Brownsnake. The closely related Red-bellied Snake normally has a bright-red belly. Ring-necked Snakes are easily confused

Dekay's Brownsnake, adult. Photograph by Allen Blake Sheldon.

with immature Dekay's Brownsnakes; however, Ring-necked Snakes have smooth body scales. Tan-colored Smooth Green-snakes can also be mistaken for Dekay's Brownsnakes.

DISTRIBUTION

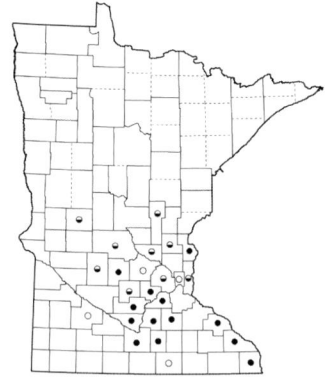

This species is found from Maine south to Florida and then west to Minnesota and Texas. The range of the Dekay's Brownsnake is essentially the eastern half of the United States. Brownsnake records in Minnesota include counties from the central and southeastern regions of the state. It is absent from the northern third and the southwest corner of the state.

HABITAT

In Minnesota, the Dekay's Brownsnake occupies a moist environment in association with deciduous forests, especially woodland edges near clearings, marshes, and ponds. Lowland forests seem more suitable than upland areas with well-drained soil. In Wisconsin, Vogt (1981) found that old fields and vacant city lots are suitable habitat. Brownsnakes hibernate in deserted ant mounds, rock piles, and stone foundations, and they are often found with gartersnakes, Red-bellied Snakes, and Smooth Greensnakes. The same hibernacula are used year after year (Minton 1972).

LIFE HISTORY

With warm weather in late April, Dekay's Brownsnakes begin to emerge from winter retreats and may be found basking on rocks. Essentially diurnal during the spring and fall, they are active during evening hours in summer months. A very secretive snake, this little serpent spends much of its time under shelter such as rocks, logs, boards, and discarded rubbish. The home range size is unknown but is expected to be small, although one individual traveled 374 meters (409 yards) in 30 days in Ontario (Freedman and Catling 1979). During the fall they may be found crossing roads on their way to hibernacula. This migration can result in high snake mortality on busy highways.

Dekay's Brownsnakes feed primarily on earthworms, slugs, and soft-bodied insect larvae. Other recorded food items for this species include snails, small frogs, amphibian eggs, small fish,

and sow bugs (Wright and Wright 1957). They actively seek prey using their sense of smell. If the prey is an earthworm, it is seized midbody. Then the snake works its mouth around to either end and swallows it whole.

By the fall of their second year, Brownsnakes ordinarily reach sexual maturity. Breeding occurs the following spring shortly after emergence. As do other snake species, females release pheromones, which are chemical odors that attract males. During August or early September, following a gestation period of 105 to 113 days, females give live birth to an average litter of 14 (range of 3 to 41) (Fitch 1970). Newborn snakes are delivered in a thin transparent membrane, from which they quickly escape. Their length at birth ranges from 9.5 to 10.5 centimeters (3¾ to 4⅛ inches).

When frightened, Dekay's Brownsnakes flatten their bodies and discharge a mild musk, but generally they cannot be provoked to bite. This species is eaten by many predators, including skunks, raccoons, weasels, opossums, shrews, hawks, robins, thrashers, toads, and other snakes.

REMARKS

According to Conant and Collins (1998), there are five recognized subspecies of the Dekay's Brownsnake; the subspecies found in Minnesota is the Texas Brownsnake (*Storeria dekayi texana*).

Red-bellied Snake

Storeria occipitomaculata

DESCRIPTION

As the name indicates, this small, secretive colubrid possesses a crimson-colored belly. Dorsal ground color may be gray, brown, reddish brown or black, and generally there is a wide, light mid-dorsal stripe or less commonly four narrow, dark stripes extending the length of the back. Usually the top of the head is darker than the back, and there may be three pale-yellow or white spots on the nape of the neck. A white chin is followed by a deep-red or orange-red belly. Dorsal scales are markedly keeled, and the anal plate is divided. Immature Red-bellied Snakes tend to be darker than adults, the spots on their neck are often fused, forming a collar, and their belly coloration is less intense.

Adults reach a total length of 20 to 25 centimeters (8 to 10 inches; Conant and Collins 1998).

The bright-red, unmarked ventral surface of the Red-bellied Snake distinguishes it from its close relative, the Dekay's Brownsnake. Juvenile Ring-necked Snakes are very similar in appearance to young Red-bellied Snakes, except Ring-necked Snakes have smooth body scales.

Red-bellied Snake, adult. Photograph by James E. Gerholdt.

DISTRIBUTION

The Red-bellied Snake ranges from southern Canada and Maine south to northern Florida and west to eastern North Dakota in the north and Texas in the south. Large gaps in the range occur in the central states. There are disjunct populations in western South Dakota and eastern Wyoming. In Minnesota, this species has a statewide distribution, although it is more common in the northern half of the state.

HABITAT

Red-bellied Snakes prefer woodland habitat; however, it is not uncommon to find them under ground cover in moist, grassy meadows. Karns (1992a) documented the occurrence of Red-bellied Snakes in peatland habitat in northern Minnesota. This species may be spotted crossing and basking on hiking trails in forested areas of the state, especially in the late summer and early fall. During spring and fall migration to winter retreats they may be found in window wells and basements. Red-bellied Snakes congregate to hibernate with other snake species in abandoned ant mounds (Lang 1969). Other possible winter retreats include rock crevices, old stone foundations, and deserted wells.

Belly of Red-bellied Snake. Photograph by Allen Blake Sheldon.

Color morphs of Red-bellied Snakes. Photograph by Minnesota Department of Natural Resources—Carol D. Hall.

LIFE HISTORY

The active season of the Red-bellied Snake in Minnesota normally extends from late April to October. This snake is thought to be diurnal in the spring and fall and crepuscular or nocturnal during the summer. Because Red-bellied Snakes are small and inconspicuously marked, they are difficult to see when crawling through leaves and twigs, and they may be more active during daylight hours than suspected. They forage in broad daylight during the summer in forest openings of northern Minnesota (Oldfield and Moriarty 1994).

Small, soft-bodied prey, such as slugs, earthworms, beetle larvae, and isopods, are preferred food items. Soft body parts of snails are extracted from their shells and consumed (Rossman and Myer 1990).

Red-bellied Snakes are sexually mature at two years of age, and courtship may take place spring, summer, or fall. Females are capable of storing viable sperm for several months before fertilization occurs. Annually, they give birth to a litter of 1 to 21 young in August or early September. Breckenridge (1944) reported that a female captured in Pine County delivered a litter of 18 young on July 17. Each newborn snake is covered with a thin, transparent membrane, which it promptly ruptures. Their length at birth ranges from 7 to 11 centimeters (2¾ to 4⅜ inches) (Ernst and Barbour 1989).

This docile species can be handled without risk. It almost never attempts to bite, and if it does, its tiny teeth are unlikely to penetrate skin. When the snake is alarmed, a mild musk is discharged from glands at the base of the tail. On several occasions Red-bellied Snakes have been reported to contort their body and stiffen out, displaying their brightly colored abdomen in a behavior similar to death feigning (Harding 1997; R. Jordon 1970). Red-bellied Snakes have also been observed creating a "grin" by flattening their head and curling their upper lip, exposing the teeth (Ernst and Barbour 1989).

A number of bird species and small predatory mammals consume Red-bellied Snakes. Milksnakes and North American Racers are also predators.

REMARKS

Of three recognized subspecies of Red-bellied Snake, two are found in Minnesota. The Northern Red-bellied Snake (*Storeria occipitomaculata occipitomaculata*) and the Black Hills Red-bellied Snake (*S. o. pahasapae*) occur in Minnesota in a broad area of intergradation across the entire state (Ernst 1974). The Northern Red-bellied Snake generally displays a light mark on the fifth upper labial scale and three well-defined light spots on the back of its neck. The Black Hills Red-bellied Snake lacks a spot on the upper labial scale, and the nape spots are missing or poorly defined.

Red-bellied Snakes may appear to be scarce in many areas; however, Lang (1969) captured and marked over 1,500 individuals on 24 hectares (59 acres) in north-central Minnesota.

Plains Gartersnake
Thamnophis radix

DESCRIPTION

The Plains Gartersnake, an occupant of Minnesota prairies, is a medium-sized striped snake. The dark-brown, dark-gray, or black ground color is partitioned by three light-colored stripes extending the length of the snake. The middorsal stripe is usually a deep yellow, and the side stripes are pale yellow with a greenish or bluish tint. The lateral stripes are located on the third and fourth scale rows above the belly plates. If the ground color is not too dark, two rows of alternating squarish spots may be discernible between the median and lateral stripe. Another row of dark spots is generally present between the belly plates and the lateral stripe. The top of the head is dark, and the pale-greenish upper lip is marked with bold, black vertical bars along the edges of the labial scales. The underside of the snake is pale green or

Plains Gartersnake, adult. Photograph by James E. Gerholdt.

dingy gray. Dorsal and lateral body scales are keeled, and the anal plate is single. Juveniles are marked similar to adults.

The total length of an adult is 38 to 71 centimeters (15 to 28 inches) (Conant and Collins 1998). Breckenridge (1944) reported an unusually large individual found in Anoka County in 1938 that measured 104 centimeters (41 inches).

Two striped snakes found in Minnesota that may be confused with the Plains Gartersnake are the Common Gartersnake and the Lined Snake. The Common Gartersnake lacks the dark bars on the upper lip, and its lateral stripes are on the second and third scale rows. Lined Snakes have a double row of half-moon spots on their abdomens.

DISTRIBUTION

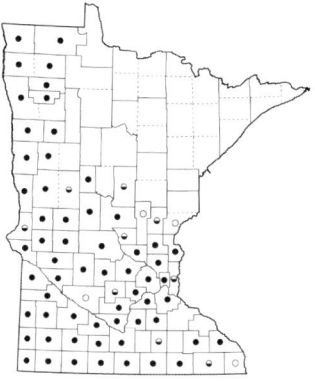

The Plains Gartersnake's range begins in western Indiana and expands westward across the Great Plains to the Rocky Mountains, reaching southern Canada in the north and northern New Mexico in the south. The distribution of the Plains Gartersnake in Minnesota includes the western, southern, and central counties. This species has not been reported from Minnesota's coniferous biome in the Arrowhead region or counties in the north-central region.

HABITAT

Plains Gartersnakes prefer open grasslands and sparsely wooded savannas. They are more frequently encountered along the edges of ponds, lakes, marshes, and streams in prairie habitats. Upland prairie distant from water is also utilized but less regularly. This species is not found in the northern coniferous forests. Abandoned mammal burrows (e.g., gopher and ground squirrel), deserted ant mounds, building foundations, and old wells are used as hibernacula (Ernst and Barbour 1989). They frequently hibernate communally with other snake species, such as Smooth Greensnakes and Red-bellied Snakes (Lang 1969).

LIFE HISTORY

If warm weather permits, the Plains Gartersnake is active from mid-April through November. An individual was observed

Amelanistic Plains Gartersnake.
Photograph by Barney Oldfield.

basing on an ant mound in Scott County on an unseasonably warm November 14 (Oldfield and Moriarty 1994). This species is principally diurnal, preferring air temperatures of 21° to 29°C (70° to 84°F). Ernst and Barbour (1989) reported observing this species feeding on Boreal Chorus Frogs at night in Minnesota. Plains Gartersnakes maintain a small, restrictive home range, moving less than 2 meters (6½ feet) per day (Seibert and Hagen 1947). The density of Plains Gartersnakes can reach 845 per hectare (340 per acre) but normally is much lower (Rossman, Ford, and Seigel 1996).

Plains Gartersnakes have a large appetite for amphibians and consume many species of frogs, salamanders, and their larvae. They also eat grasshoppers, beetles, earthworms, slugs, fish, and small mammals (Ernst and Barbour 1989; O. R. Jordan 1967). Rodents may become a significant portion of their diet in areas distant from standing water.

Breeding normally occurs from mid-April through May, although Collins (1982) reported fall activity. Often several males court a single female by gliding alongside and nudging her with their noses while continually flicking their tongues. The male that gains the best position makes cloacal contact and achieves copulation while the receptive female remains stationary and elevates her tail. After a successful breeding, the male leaves a seminal plug in the cloaca of the female to inhibit copulation with other males (Ross and Crews 1977). Females give live birth to litters averaging 25 in size during late August and September. Litter size varies tremendously, ranging from 5 to 97 (Smith and Zimmer 2013). Newborns range in length from 15 to 23.5 centimeters (5⅞ to 9¼ inches), and they may grow to 45 centimeters (17¾ inches) by the end of their first year.

When threatened, Plains Gartersnakes may hide their head under body coils and slowly wave their tail tip as a decoy (Oldfield and Moriarty 1994). If picked up, this species struggles to escape and sprays a pungent musk from glands near the base of the tail. It will attempt to bite but is not as aggressive as the Common Gartersnake.

Raptors and carnivorous mammals are the most common predators and include hawks, American kestrels, red foxes, coyotes, striped skunks, and raccoons (Ernst and Barbour 1989). Habitat destruction and vehicles take a greater toll on populations than natural predators do.

REMARKS

The two subspecies listed in Oldfield and Moriarty (1994) are no longer considered distinct subspecies, and there are currently no recognized subspecies for the Plains Gartersnake (Crother 2012).

Common Gartersnake

Thamnophis sirtalis

DESCRIPTION

The most commonly encountered snake in Minnesota and the only species that many Minnesotans ever see in the wild is the Common Gartersnake. This harmless species is medium sized with a slender build and three longitudinal yellow stripes that may be tinted blue, green, brown, or orange. The median stripe is generally a lighter hue than the lateral stripes. The lateral stripes are located on the second and third scale rows, counting up from the belly plates. The ground color is black, grayish brown, or olive. A double row of black spots may be apparent between the dorsal and lateral stripes. Common Gartersnakes from southwestern Minnesota normally have prominent red bars along their sides between the stripes. Usually these bars are much reduced or absent on individuals from eastern and northern Minnesota. The top of the snake's head is dark, and the unmarked upper lips are light yellow or pale green. The chin is off-white, and the pale-colored belly is yellow, green, or blue. Dark spots may be present along the outside margins of the ventral plates.

Common Gartersnake.
Photograph by Tony Gamble.

Body scales are keeled, and the anal plate is undivided. Young are marked similar to adults.

The average adult Common Gartersnake is 41 to 66 centimeters (16 to 26 inches) in length (Conant and Collins 1998). An exceptionally large female from Fillmore County measured 104 centimeters (41 inches).

The similar-looking Plains Gartersnake has bold lip bars, and its lateral stripe is found on scale rows three and four. The Lined Snake has black half-moon spots on its belly.

DISTRIBUTION

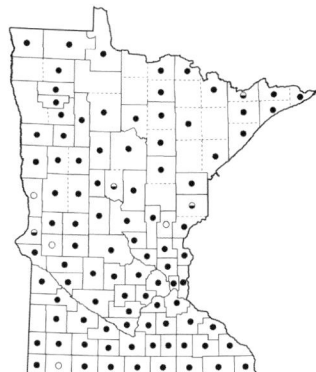

The Common Gartersnake has the widest distribution of any North American snake and lives farther north in Canada than any other serpent. This species is found from the East Coast to the West Coast and south to Florida and eastern Texas. It is not found in large areas of the arid Southwest. The Common Gartersnake has a statewide distribution in Minnesota.

HABITAT

The Common Gartersnake is a habitat generalist and resides in almost any vegetation type found in Minnesota, including deciduous and coniferous forests, marshes, peatlands, and prairies. This species tolerates some degree of habitat alteration, as it is found on farmsteads, urban lots, golf courses, and cemeteries. It is strongly affiliated with water and is often found near ponds, streams, sloughs, and swampy areas.

Suitable hibernacula include mammal burrows, building foundations, rock crevices in bluffs and quarries, and deserted ant mounds (Ernst and Barbour 1989). Large numbers of this species may hibernate together. They are regularly found around building foundations and concrete front stoops in the outer suburbs (Moriarty 1999). Congregations of approximately 8,000 snakes have been reported in limestone sinkholes in Canada (Gregory 1977).

LIFE HISTORY

Being relatively cold tolerant, the Common Gartersnake may be out basking shortly after a snow melt in early April and has

Melanistic Common Gartersnake. Photograph by Jeffrey B. LeClere.

Red-sided Gartersnake, adult. Photograph by Allen Blake Sheldon.

Common Gartersnake, juvenile
and recently fed. Photograph by
Tony Gamble.

been observed into December (T. Jessen, pers. comm.). The species is considered to be primarily diurnal; however, it may be active during the evening hours in spring and summer. Individuals may search for frogs and toads at night during amphibian breeding congregations (Ernst and Barbour 1989). According to Aleksiuk (1976), active snakes have a body temperature range of 18° to 30°C (64° to 86°F). Common Gartersnakes may wander far afield to find summer feeding areas. Movements as great as 17.7 kilometers (11 miles) have been reported (Gregory and Stewart 1975).

Common Gartersnakes are not finicky eaters, and the list of acceptable food is extensive. Amphibians and earthworms are principal food items. Other food staples include fish, nestling birds, small mammals, arthropods, mollusks, and carrion. On rare occasions, this species has been known to eat young snakes of other species (Ernst and Barbour 1989). Prey is trailed by scent and is seized by mouth and swallowed. Although the bite of the Common Gartersnake is harmless to humans, there is some evidence that its saliva contains enzymes that help subdue and immobilize a struggling victim (Ernst and Barbour 1989).

Clusters of courting Common Gartersnakes can be found near hibernacula in early spring. These writhing congregations are typically composed of a single female (usually the largest of the group) and a number of males. Each male attempts to

Breeding mass of Common Gartersnake. Photograph by Barney Oldfield.

maneuver into a position that will enable him to achieve cop- ulation. The actual mating may last 15 to 20 minutes. The male leaves behind a copulatory plug in the female's cloaca that inhib- its successful mating by other males. Gibson and Falls (1975) determined that despite the copulatory plug, a female may en- gage in multiple breedings. The female is capable of storing live sperm through the winter for fertilization of eggs the following spring. Females mature at two years of age and reproduce annu- ally in Minnesota. Litter size ranges from 3 to 103 (Wright and Wright 1957), although 10 to 25 is typical. Newborns are 13 to 23 centimeters (5 to 9 inches) long.

When alarmed, the Common Gartersnake makes a dash for protective cover. This species is primarily terrestrial but readily slides into nearby water to escape predators. If cornered, they ei- ther attempt to hide their head under body coils, or they flatten their neck and strike at the intruder. When captured, they strug- gle violently, bite repeatedly, discharge cloacal contents, and spray a pungent musk. The musk of this species is strong and lingers on the captor's skin. If a person is bitten, the wound generally bleeds, but it causes no more discomfort than a minor scratch.

Common Gartersnakes have many predators, including mammals, birds, reptiles, amphibians, and fish. Loss of habitat, busy roads, and pollution take a heavy toll on these snakes.

REMARKS

A broad zone of intergradation for two subspecies of the Common Gartersnake occurs across Minnesota, extending from the northwest corner to the southeast corner (Conant and Collins 1998). The eastern part of the state is inhabited by the Eastern Gartersnake (*Thamnophis sirtalis sirtalis*), which generally lacks any red coloration. If there is red coloration, it occurs in small specks on the skin between scales along the snake's sides. As a rule, the Red-sided Gartersnake (*T. s. parietalis*) has bold red blotches along its sides between the median and lateral stripes, and it is more likely to be encountered in the western part of the state.

People commonly refer to the Common Gartersnake as a "grass snake" or "gardener snake," probably due to the frequency that it is found in gardens or yards.

Lined Snake

Tropidoclonion lineatum

DESCRIPTION

The Lined Snake is a small, striped snake of the prairies. Dorsal coloration is a drab gray brown, olive gray, or dark gray with three dirty-white or pale-yellow stripes running the length of the body. The middorsal stripe is narrow, and the lateral stripe occupies scale rows two and three. There may be a row of small, black spots running alongside the middorsal stripe as well as another row following along the dorsal edge of the lateral stripe. The top of the small head has dark stippling. The ventral surface is white or pale yellow and is distinctly marked with a double row of black or near-black solid semicircles. Dorsal body scales are keeled, and the anal plate is single. Juvenile Lined Snakes have poorly defined stripes, but the half-moon belly spots are distinct.

Lined Snake, adult. Photograph by Barney Oldfield.

Belly of Lined Snake. Photograph by Barney Oldfield.

The total length of adult Lined Snakes ranges from 22 to 38 centimeters (8¾ to 15 inches) (Conant and Collins 1998). There are no maximum measurement records for Minnesota, as very few individuals have been found.

The larger gartersnakes of Minnesota have a similar striped pattern as the Lined Snake, but gartersnakes lack the double row of black spots on their abdomens.

DISTRIBUTION

The fragmented distribution of the Lined Snake begins in extreme southeastern South Dakota, extends into central Iowa, and covers the Great Plains southward into central Texas. Isolated populations occur in Illinois, Iowa, Missouri, Colorado, and New Mexico. The only records for the Lined Snake in Minnesota are in Rock County. This species is found across the border in South Dakota in Minnehaha County.

Lined Snake habitat in southwest Minnesota. Photograph by Minnesota Department of Natural Resources—Carol D. Hall.

HABITAT

Lined Snakes prefer prairies, woodland edges, and open areas with scattered trees. In Minnesota its habitat is rolling prairie with large amounts of exposed rock. Elsewhere, they live in vacant city lots, cemeteries, trash dumps, and along highways where there is abundant ground cover (T. R. Johnson 2000). Little is known about hibernation sites. Hamilton (1947) found seven Lined Snakes hibernating together in Texas. They were coiled 7.5 to 20 centimeters (3 to 8 inches) deep in the soil with their heads protected in the center of their body coils. They evidently seek shelter below the frost line in rock crevices or burrows.

LIFE HISTORY

Lined Snakes emerge in late April and return to their winter retreats by October. They may bask in the open during the day in the spring and again in early fall; however, this species is normally nocturnal, spending daylight hours hidden under rocks, logs, bark, and other debris.

Lined Snakes feed almost exclusively on fossorial invertebrates, locating them by smell and taste at night when they come out on the surface of the ground. Sow bugs are also eaten. Since this snake has a small head and mouth, its prey is restricted to appropriately sized soft-bodied animals (Ernst and Barbour 1989).

Females mature in two years and may give birth annually. Mating generally takes place in the fall, and the female retains viable sperm until the following spring, when fertilization occurs (Fox 1956). During the month of August, litters of 2 to 13 young are born with a range in length of 7 to 13 centimeters (2¾ to 5⅛ inches). The young are born in thin, transparent membranes, from which they free themselves.

When threatened by an intruder, this small snake hides its head under its body. It often voids cloacal contents and sprays musk if disturbed, but is very unlikely to bite. Even if it tried to bite, the resultant injury to a human finger would be inconsequential. Striped skunks, raccoons, weasels, and a variety of birds are known to eat these snakes (Ernst and Barbour 1989). Road-killed specimens have been found within and surrounding Blue Mounds State Park (T. Jessen, pers. comm.).

REMARKS

Currently, there are no recognized subspecies of the Lined Snake (Crother 2012). The Lined Snake was first discovered in Minnesota in 1972. Due to its restricted range within the state, the Lined Snake is classified as special concern and listed as a Species of Greatest Conservation Need by the Minnesota DNR (2006, 2013). The population within Blue Mounds State Park is protected by law due to their location in the park.

Family Viperidae—Pit Vipers

Of the 278 species of venomous vipers found worldwide, only two species are found in Minnesota. The two viper species in Minnesota are in the subfamily Crotalinae (pit vipers). Pit vipers have unique, highly sophisticated, heat-sensitive facial pits located on each side of the head between the eye and nostril. These pits allow the snake to locate warm prey even in total darkness. All pit vipers have vertical pupils. They have stout bodies, narrow necks, broad heads, and a pair of hollow retractable fangs on the front of the upper jaw. The United States is home to 19 species of pit vipers, 15 of which are members of two genera of rattlesnakes. Even though many species of snakes (venomous and nonvenomous) vibrate their tail when alarmed, only rattlesnakes have a segmented rattle on the end of their tail that produces a distinctive sonorous buzz.

Pit vipers found in Minnesota are the Timber Rattlesnake and the potentially extirpated Massasauga. These snakes are dangerous to humans, and they should be avoided or observed from a safe distance.

Timber Rattlesnake
Crotalus horridus

DESCRIPTION

The Timber Rattlesnake is a large, heavy-bodied rattlesnake native to the bluff and hill country of southeast Minnesota. This snake has a broad head, distinct from its narrow neck, and a large, tan rattle on the end of its black tail. Dorsal ground color may be yellow, tan, brown, reddish brown, or, rarely, gray. On some individuals the dark coloration of the tail extends to include the posterior third of the body. Black or near-black, chevron-like cross bands form a pattern along the back, but several of these bands may be incomplete or broken on the front third of the body. Most snakes have a single row of light-colored scales bordering the dark cross bands. There may be a rust-colored middorsal stripe from the back of the head to the tail or even to the base of the rattle. In general the top of the head is the same ground color as the body and is unmarked except for a slightly darker nose. Frequently, a dark stripe extends from the eye to the angle of the mouth. The lips and chin are yellow or tan, and the belly is yellowish tan or light gray with a scattered stippling

Timber Rattlesnake, adult.
Photograph by Barney Oldfield.

of dark gray or black that becomes heavier toward the tail. Dorsal body scales are strongly keeled, and the anal plate is single. There is a sensory pit between the eye and nostril on either side of the face, and the eyes have elliptical pupils in daylight.

Juvenile Timber Rattlesnakes are patterned like adults, but they tend to have a pinkish-tan ground color and a prominent, reddish middorsal stripe. Newborns are gray with black bands until their first skin shedding, which occurs at 10 to 14 days of age.

Excluding the rattle, adult Timber Rattlesnakes range in length from 80 to 122 centimeters (31½ to 48 inches). Adult females average 90 centimeters (35½ inches), and adult males average 110 centimeters (43¼ inches; Keyler and Oldfield 1992). Snakes that exceed 122 centimeters (48 inches) are rare in Minnesota. The largest known individual from Minnesota was a male from Houston County that measured 135 centimeters (53 inches) and weighed 1,760 grams (3 pounds, 14 ounces; Keyler and Oldfield 1992).

In contrast to the Timber Rattlesnake, the Massasauga is a significantly smaller, blotched rattlesnake without a distinctly darker tail. Several species of harmless colubrids (including the Western Foxsnake, Gophersnake, Common Watersnake, and Milksnake) are regularly confused with the Timber Rattlesnake. None of these colubrids have facial pits, elliptical pupils, or a rattle.

DISTRIBUTION

Much of the former range of the Timber Rattlesnake has been reduced by habitat loss and persecution. The snake was once found over much of the eastern United States excluding Michigan, upstate Maine, and the Florida peninsula. There are records for Timber Rattlesnakes in eight of Minnesota's southeastern counties; however, the likelihood of current viable populations in Washington or Dakota Counties is remote. Currently, the most northerly active dens are known from Goodhue County, and they are widely scattered. Historical (late 1800s) newspaper stories on snakebites can be found in Blue Earth and Stearns Counties (Mahaffy 2009). These reports cannot be verified.

Newborn Timber Rattlesnake. Photograph by Allen Blake Sheldon.

HABITAT

In Minnesota, the Timber Rattlesnake is found in the steeply dissected hills and valleys of the Mississippi River drainage. Forested hillsides with south- to southwest-facing rock outcrops and bluff prairies are critical habitat components. The snakes use limestone, dolomite, or sandstone bluffs and outcrops for hibernacula. Nearby forests, forest edges, and occasionally croplands serve as summer feeding grounds. This species hibernates in ancestral communal dens in rock fissures and crevices. Other snake species such as North American Racers, Western Ratsnakes, Milksnakes, and Gophersnakes may share hibernacula with the Timber Rattlesnake.

LIFE HISTORY

Timber Rattlesnakes occasionally bask near den openings in late April with favorable weather, but May is the prime emergence month in Minnesota. According to Brown (1982), New York populations disperse by mid-June, and individuals travel an average of 504 meters (551 yards) to their summer foraging areas. The maximum recorded distance of travel from the den is 5.6 kilometers (3.5 miles). They are active during the day in the spring and fall but become primarily nocturnal during summer months.

Even though Timber Rattlesnakes are basically terrestrial, they are adept at climbing along rocky ledges and scaling bluff walls. Occasionally they are found in trees or vines up to 2 meters (6½ feet) off the ground (Oldfield and Keyler 1989). They are also capable swimmers if the need arises.

Small mammals are their preferred prey, which includes mice, shrews, moles, rats, chipmunks, squirrels, and young rabbits. Occasionally small birds, insects, and amphibians are also eaten (Wright and Wright 1957; Klauber 1972). A study in Pennsylvania found the Timber Rattlesnake to be a "sit-and-wait" ambush predator (Reinert, Cundall, and Bushar 1984). A hungry snake spends several hours during the night coiled alongside a log waiting for an unsuspecting meal. Downed logs serve as runways for rodents. When prey comes into range, the

Rattlesnake rookery. Photograph by James E. Gerholdt.

snake's heat-sensing pits direct the strike. The rodent runs a short distance and quickly succumbs to the effects of the venom. The snake follows the scent of the victim, and after catching up to it, swallows it whole.

Timber Rattlesnakes mature late, as females are 5 to 6 years old before they give birth to their first litter (Martin 1966). Brown (1991) found females in New York State requiring 7 to 11 years to reach maturity. Sexually mature males occasionally engage in a combat dance in April or May (Collins 1974; Klauber 1972). This rarely witnessed behavior involves two adult males elevating the front third of their bodies off the ground and attempting to push or throw their rival to the ground to establish dominance. Females reproduce every 3 or 4 years (Brown 1991). Courtship and mating occur in late summer, and the female stores sperm over the winter for egg fertilization the following spring. Pregnant females remain within proximity of the den, taking advantage of open areas for periods of basking to hasten embryo development. A litter of 3 to 17 young (average of 7) is born in late August or September. Newborn Timber Rattlesnakes are 20 to 36 centimeters (8 to 14 inches) in length, and they remain with their mother for the first 10 to 14 days.

At birth they have a prebutton on the end of their tail, which gives way to a button (terminal segment of a complete rattle) when the skin is shed the first time. They gain a rattle segment with each subsequent shedding. (Mature snakes shed once or twice a year, and young snakes shed more frequently.) Since rattles are often missing segments, they do not provide a reliable method for determining the age of the snake.

Typically, the Timber Rattlesnake is a secretive, shy animal that remains motionless, depending on cryptic coloration to avoid detection. If discovered, the snake often makes a rapid exit to protective cover and buzzes its rattle. When caught out in the open without a safe escape route, it will coil, flatten its body, and buzz continually. If harassed, it will either hide its head under body coils or launch a strike in the direction of the intruder.

The bite of the Timber Rattlesnake is dangerous and has resulted in human fatalities (Keyler 1983, 2005), but with prompt medical attention the prospect of death is unlikely. The symptoms of a bite include extreme pain, swelling, weakness, nausea, vomiting, difficult breathing, reduced blood pressure, and unconsciousness in severe cases. Available case histories indicate that approximately 25 percent of Timber Rattlesnake bites are "dry" bites in which no venom is injected (Keyler 2005).

Juvenile snakes have many predators, including hawks, owls, carnivorous mammals, and other snakes such as Racers and Milksnakes. Humans are the dominant predators of adults, and they have contributed greatly to the decline of Timber Rattlesnakes.

REMARKS

Additional common names for the Timber Rattlesnake include "ol' velvet-tail," "banded rattlesnake," and "bluff rattler." This species has been reported to survive just over 30 years in captivity (Bowler 1977).

Historically, dens supported large populations of Timber Rattlesnakes, but due to years of persecution by snake hunters, very few sizable dens remain. Many are entirely depleted of snakes. Bounty was paid by local governments to rattlesnake hunters in Minnesota until 1989, when a bill to repeal the bounty was signed into law. The Timber Rattlesnake is classified as threatened and is listed as a Species of Greatest Conservation Need by the Minnesota DNR (2006, 2013). Timber Rattlesnakes are fully protected in Minnesota. It is illegal to harass, kill, or collect them in the state. The DNR has developed a Timber Rattlesnake Recovery Plan (Timber Rattlesnake Recovery Team 2008). Conservation efforts focusing on habitat improvements and public outreach have helped to change attitudes regarding this threatened species.

Massasauga

Sistrurus catenatus

DESCRIPTION

The Massasauga is a small- to medium-sized, robust, spotted rattlesnake of river-bottom floodplains. Ground color ranges from light gray to medium brown. A row of 26 to 40 large, dark-brown or near-black blotches extends down the back. The blotches are often outlined with a thin, light line. Two or three additional rows of smaller, alternating spots are found along each side of the body. The row nearest the belly blends into the dark, almost solid black abdominal scales that are lightly mottled with white or yellow. The snake's tail is ringed with large, dark bands of the same color as the dorsal spots. There are two broad, dark stripes that stretch from the top of the head out onto the neck. An additional dark stripe with a distinct white border goes from the eye near the jaw angle to the neck on each side of the face. The chin

Massasauga, adult female. Photograph by Allen Blake Sheldon.

is dirty white. A heat-sensitive pit is located on either side of the face between the nostril and the eye. The vertical pupils form thin, black lines in bright sunlight. Males have a proportionally longer tail than females, and the tail ends with a cornified rattle. Body scales are keeled, and the anal plate is undivided. Juvenile Massasaugas are patterned as adults but with lighter ground color, and the area between the bands on the tail is bright yellow.

Total length of this species seldom exceeds 91 centimeters (36 inches) excluding the rattle. Most adults are 47 to 76 centimeters (18½ to 30 inches; Conant and Collins 1998).

All nonvenomous species in Minnesota that are likely to be confused with the Massasauga lack rattles, facial pits, and vertical pupils. The only other snake in Minnesota with rattles and a vertical pupil is the larger Timber Rattlesnake, and it has cross bands and a solid black tail.

DISTRIBUTION

The range of the Massasauga in North America begins in southern Ontario and central New York and extends obliquely through the eastern deciduous forest onto the Great Plains into south Texas, New Mexico, and southeastern Arizona. There are disjunct populations in several states. Only Wabasha and Houston Counties in Minnesota have Massasauga records, and they have been notably few. Breckenridge (1944) mentioned only two specimens, both from Wabasha County. A scattering of unconfirmed sightings has occurred during the past 50 years. Levell (1994) raised doubts about the validity of the two specimens and other records. Mahaffy (2009) reports historical newspaper articles from the 1800s on snakebites that suggest Massasaugas may have been found in additional counties.

HABITAT

The Massasauga occupies moist habitats such as marshes, bogs, swamps, and associated wetlands. In Wisconsin, Massasaugas are restricted to river-bottom lowland forests and adjacent open fields (Vogt 1981). Massasaugas apparently do not den with other snakes during the winter. They hibernate in mammal burrows, old tree stumps, crayfish burrows, and rocky crevices.

Newborn Massasauga.
Photograph by Allen Blake
Sheldon.

LIFE HISTORY

Prompted by spring flooding, Massasaugas emerge in late April in Wisconsin (Vogt 1981). During times of high water they may be found on muskrat houses or other partially submerged brush piles and trees. As the water recedes, they move to meadows and open fields within the bottomlands. Reinert and Kodrich (1982) determined the mean home range for Massasaugas in Pennsylvania to be slightly less than 1 hectare (2½ acres). They are primarily diurnal during the spring and fall but resort to crepuscular and nocturnal activity during the summer. Gravid females are fairly sedentary with small home ranges (Marshall, Manning, and Kinsbury 2006) and can be found basking on humid, overcast days in late summer (Oldfield and Moriarty 1994). By mid-October they return to their hibernating sites to escape freezing temperatures.

Adult Massasaugas feed chiefly on small rodents, such as voles, mice, and shrews. The heat-sensitive pits are used to detect endothermic prey, but sight and odor are also important stimuli. Normally, the prey is consumed after it dies from the effects of envenomation. Subadults feed on frogs, large insects, Dekay's Brownsnakes, and juvenile Common Gartersnakes. The bright-yellow coloration on the tail of a young Massasauga serves as a lure for frogs that often feed on moving, brightly colored insects (Schuett, Clark, and Kraus 1984).

Females become reproductively mature by their third year, and most breeding activity takes place in the spring, although fall courtship has been reported. The Massasauga may not reproduce annually. Reinert (1981) in Pennsylvania and Seigel (1986) in Missouri have found evidence of a biennial reproductive cycle. The gravid female gives birth to 3 to 20 young during late August or early September. Newborns range in length from 18 to 25 centimeters (7 to 10 inches) and have a small prebutton on the end of their tail. Young shed their skin at five days of age, and the prebutton is replaced with a button (the terminal segment of a complete rattle). After several days of staying near their mother, young snakes begin life on their own (Oldfield and Moriarty 1994).

Massasaugas generally remain motionless, depending on cryptic coloration to prevent detection, but nervous individuals give their location away by buzzing their rattle. Without an escape route they hold their ground, vibrate their tail, and strike if harassed. Massasauga venom is potent, but due to the snake's modest size, relatively short fangs, and small venom glands, the bite is generally not life threatening to humans.

Hawks, large wading birds, and small carnivorous mammals are natural predators, but persecution of the Massasauga and destruction of its habitat are the causes of a serious population decline across the United States.

REMARKS

Three subspecies of the Massasauga are described, but only the Eastern Massasauga (*Sistrurus catenatus catenatus*) is found in Minnesota (Conant and Collins 1998).

The Massasauga has several local names, including "swamp rattler," "black snapper," and "sauga." According to Vogt (1981), the name *massasauga* is a Chippewa Indian word meaning "great river mouth." Captive longevity for this species is 14 years (Bowler 1977). The Massasauga is classified as endangered and is listed as a Species of Greatest Conservation Need by the Minnesota DNR (2006, 2013).

The current status of the Massasauga in Minnesota is in

doubt. Numerous herpetologists have searched for them in Minnesota to no avail (Levell 1994; Naber, Majeski, and DeMars 2004). The few records that exist are all from or adjacent to the main channel of the Mississippi River, downstream of Wisconsin's Chippewa River. There is a known Massasauga population in that drainage (Vogt 1981). Levell (1994) suggests that individuals raft across the Mississippi River on logs or debris from the Chippewa River. The Massasauga should be considered extirpated from the state since there have been no verified reports or sightings in the past 25 plus years.

Family Chelydridae— Snapping Turtles

Only four species are members of this turtle family, two of which occur in the United States. Snapping turtles have massive heads, small X-shaped plastrons, and a long tail armed with a crest of bony plates. These turtles are highly aquatic and are seldom seen on land except during periods of nesting.

The largest turtle in Minnesota and the sole representative of Chelydridae is the Snapping Turtle. It is found in almost any permanent body of water across the state.

Snapping Turtle

Chelydra serpentina

DESCRIPTION

Minnesota's largest turtle is well known for its ill disposition and aggressive attitude. The Snapping Turtle is equipped with a massive carapace and a large head with powerful jaws. The carapace has three longitudinal keels, which wear down and become smooth in old adults. Serrations are apparent on the back edge of the dorsal shell. The tail is longer than the carapace and has a dorsal row of sawtooth projections. The carapace, which is often covered with mud or algae, is brown, olive, or black. Normally there is a pattern of radiating lines on each scute in young turtles. Little protection is provided for the underside of the turtle by the small, X-shaped plastron, which is light yellow or tan. The dorsal surface of the limbs and head is gray or brown with varying amounts of mottling. The upper and lower jaws of the mouth are pale yellow with vertical, black lines. Tubercles are present on the long neck. Snapping Turtles have well-developed limbs with strong claws.

Male Snapping Turtles reach larger sizes than females, and

Snapping Turtle, adult. Photograph by Allen Blake Sheldon.

their tail is proportionally longer. Hatchling Snapping Turtles are dark brown to black and bear small, white flecks on their underside.

Straight-line measurements of the carapace of average adult Snapping Turtles range from 20.3 to 36 centimeters (8 to 14 inches), and the average adult weighs 4.5 to 16 kilograms (10 to 35 pounds) (Conant and Collins 1998). The largest known Minnesota individual was a male snagged by a fisherman in the Popple River in Itasca County during the summer of 1986 (Gerholdt and Oldfield 1987). "Minnesota Fats" had a carapace length of 49.4 centimeters (19½ inches) and weighed 29.6 kilograms (65 pounds).

DISTRIBUTION

The geographic range of the Snapping Turtle is from Maine and adjacent Canada west to Montana and south to New Mexico and southern Florida. This species is found across the entire United States east of the Rocky Mountains. The Snapping Turtle has a statewide distribution in Minnesota.

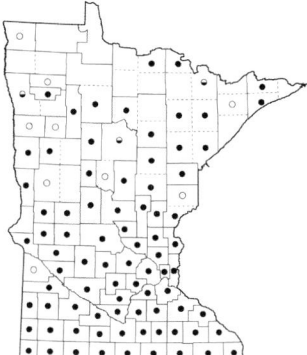

HABITAT

Snapping Turtles make a home in almost any type of permanent freshwater body, including ponds, lakes, marshes, rivers, creeks, and backwater sloughs. Larger populations of Snapping Turtles are found in water bodies with mud bottoms and abundant aquatic vegetation.

These turtles often congregate in large numbers to hibernate below the ice in muskrat tunnels, streams, and holes in riverbanks (Vogt 1981). They may bury themselves in decaying vegetation and mud or wedge under submerged logs while waiting out freezing weather (Steyermark, Finkler, and Brooks 2008). On occasion, Snapping Turtles have been seen slowly moving through the water under the ice during midwinter (Ernst and Lovich 2009; Steyermark, Finkler, and Brooks 2008).

LIFE HISTORY

Snapping Turtles become active with the onset of warm weather in late April or May and can be seen basking with other turtle

Snapping Turtle basking.
Photograph by Minnesota
Department of Natural
Resources—Carol D. Hall.

species. They bask less frequently during the summer because they are intolerant to high temperatures and can rapidly become dehydrated (Ernst and Lovich 2009).

Snapping Turtles are highly aquatic. They either lie on the bottom in shallow water, extending their long neck just far enough to project their nostrils above the water for air, or they float beneath the water surface with nostrils and eyes exposed. During the day they frequently hide beneath submerged logs, rocks, and debris. When night arrives, they become active and crawl along the bottom in search of food. Overland migration of adults occasionally occurs during wet springs (Oldfield and Moriarty 1994). If their pond dries up during summer, they are forced to make an overland trek in search of water.

The omnivorous Snapping Turtle is not a finicky eater. It eats insects, crayfish, clams, snails, earthworms, leeches, freshwater sponges, fish, fish eggs, frogs, toads, tadpoles, amphibian eggs, salamanders, snakes, small turtles, birds, small mammals, carrion, and various types of aquatic plants (Ernst and Lovich 2009). An unusual event was reported in Minnesota in which a

Snapping Turtle killed an adult trumpeter swan. The submerged turtle grasped the head of the feeding swan in its mouth, and the bird drowned while attempting to escape (Moriarty 1990). Snapping Turtles must feed while submerged because water pressure is essential for swallowing. Small food items are swallowed whole, and larger ones are held with the mouth and torn apart with their claws. Adult turtles tend to wait in ambush for prey, while young Snapping Turtles are more likely to forage for food.

Snapping Turtles are sexually mature at a carapace length of 20 centimeters (7⅞ inches). Christiansen and Burken (1979) reported that in Iowa it took four to five years for males, and six to seven years for females to mature. Maturity takes longer in the northern part of their range. Mating occurs during chance encounters anytime from April to October. Legler (1955) described underwater courtship activity between a pair in which they faced each other on the bottom and made sideward sweeps of their necks in sequences of 10 at intervals of 10 seconds. Actual mating takes place when the male mounts the female and forcibly grips her carapace with his claws, while he bites at her head and neck. Copulation is achieved when he curls his tail under hers, inserting his penis into her cloacal opening. Female Snapping Turtles are known to retain viable sperm for several years (Steyermark, Finkler, and Brooks 2008).

Egg laying takes place during May and June in Minnesota. Females often make lengthy trips away from water to seek suitable open areas for nesting. Sandy riverbanks, open fields, road embankments, muskrat houses, and lawns are commonly

Snapping Turtle, hatchling. Photograph by Allen Blake Sheldon.

Snapping Turtles emerging from nest. Photograph by Tom Jessen.

chosen. The greatest incidence of nesting activity occurs in early morning or late evening when temperatures are greater than 15.5°C (60°F) and there is a light rain (Hammer 1969). The nest is dug entirely with the female's hind legs to a depth of 10 to 17.8 centimeters (4 to 7 inches). At one-minute intervals she typically deposits 20 to 30 eggs (can range up to 83), guiding them into the cavity with her hind legs. The white eggs have a tough, leathery shell and are shaped like Ping-Pong balls with a diameter of 2.3 to 3.3 centimeters (⅞ to 1¼ inches). After covering the nest, she returns to the water. Egg development takes 55 to 125 days depending on weather, but most eggs hatch in approximately 60 days. Hatchlings have a carapace length of 2.6 to 3.1 centimeters (1 to 1¼ inches). Rhen and Lang's (1998) study at Itasca State Park showed eggs incubated at warmer temperatures produced more females and grew faster than turtles incubated at cooler temperatures.

Snapping Turtles pulled from their watery home or cornered on land live up to their name by repeatedly lunging and striking with their mouth agape. They are capable of delivering a painful and damaging bite, so care should be exercised when handling them. A Snapping Turtle is best carried by securely grasping the turtle's hind legs near the shell and holding the turtle at a safe distance from one's legs. Suspending a large Snapping Turtle by its tail can cause serious damage to the turtle's spinal cord. Typically, agitated turtles emit a strong, pungent odor. While in the

Snapping Turtle plastron. Photograph by Tony Gamble.

water, they seem more intent on crawling or swimming away than standing their ground.

Adult Snapping Turtles have few predators other than humans and their automobiles, but natural predation of eggs and young is remarkably high. Raccoons, skunks, foxes, and opossums dig up nests and eat freshly laid eggs. Nest destruction rates in some areas approach 90 percent (Harding 1997). Hatchlings and immature turtles fall prey to crows, herons, bitterns, American Bullfrogs, snakes, and large predatory fish.

REMARKS

"Snapper" is a widespread nickname.

Snapping Turtles are a major commercial turtle species in Minnesota for human consumption. During the winter, commercial turtle hunters locate hibernating congregations by probing through the ice using long poles. The turtles are pulled up through the ice with a sharp hook on the end of the pole. Baited turtle traps are commonly used during warmer months. Commercial turtle trappers reported harvesting 823 in 2010, and 813 in 2011 (Larson 2012). Turtles are protected from harvest during the nesting season.

A study by the Minnesota Pollution Control Agency in 1982 showed that Snapping Turtles from the Mississippi River in the southeastern part of the state contained high levels of toxic PCBs (Helwig and Hora 1983), bioaccumulated throughout these long-lived turtles' lives. This report should make people think twice before eating these turtles.

Family Emydidae—
Pond and River Turtles

Emydidae is a large family of turtles with 48 species and 12 genera. Excluding Australia, they are found worldwide in tropical and temperate regions and have the highest diversity in North America (33 species, 11 genera). Many members of this family are highly aquatic and spend much of their time in or close to water. These small- to medium-sized turtles have a well-developed carapace and plastron with a broad bridge connecting the upper and lower shells on either side. There is some degree of webbing between the toes to aid swimming activities. Many species also exhibit elaborate skin patterns in bright colors of yellow, red, or green.

Minnesota is home to seven species of pond and river turtles. These are the Painted Turtle, Blanding's Turtle, Wood Turtle, Northern Map Turtle, Southern Map Turtle, False Map Turtle, and Pond Slider.

Painted Turtle

Chrysemys picta

DESCRIPTION

The well-known, small- to medium-sized Painted Turtle is appropriately named for the colorful design on its plastron and marginal scutes. Its oval carapace is smooth, flattened, and unkeeled. The upper shell is dark olive to black with a net-like pattern of faint lines. The reddish color on the edge of the carapace continues onto the ventral surface of the marginal scales, interrupted by patterns of black and yellow. The large plastron is heavily bridged to the carapace and has a ground color of pale to bright orange or red with an elaborate design of black, gray, tan, and yellow. The head, legs, and tail are dark olive or black with conspicuous yellow or red lines. Generally, a bold yellow streak is present behind each eye. The upper jaw is notched.

Mature male Painted Turtles are smaller than females. Males also have a more oblong-shaped carapace, elongated fore claws, and a longer tail with the vent beyond the carapace edge. Hatchling Painted Turtles have a circular carapace with a weak keel. Their coloration and markings are similar to adults', although colors found on the plastron tend to be more vivid in

Painted Turtle, adult, basking. Photograph by Allen Blake Sheldon.

young turtles. The caruncle (egg tooth) drops off by the fifth day posthatching (Vogt 1981).

Adult Painted Turtles have a carapace length of 9 to 18 centimeters (3½ to 7 inches) (Conant and Collins 1998). The largest recorded Minnesota individual was a female from Murray County with a carapace length of 21.5 centimeters (8½ inches) (Oldfield and Moriarty 1994).

DISTRIBUTION

Painted Turtles are found from Maine south to northern Georgia and west to the eastern border of Texas. The northern border of their range runs west through southern Canada and the northern United States to Washington and Oregon. Scattered, disjunct populations of Painted Turtles are found in several southwestern states. This species is the only conspicuous basking turtle in many northern states (Conant and Collins 1998). In Minnesota, the Painted Turtle has a statewide distribution, although it is less common in the lakes of the Arrowhead Region.

HABITAT

Permanent bodies of water (such as ponds, lakes, marshes, sloughs, and creeks) with soft bottoms, aquatic vegetation, and ample basking sites are the preferred habitat of the Painted Turtle (Ernst and Lovich 2009). Individuals are also found in the state's larger rivers and lakes with rocky shores, but these localities do not support large populations.

Hibernating occurs in water and mud under the ice. Large numbers may group together against submerged logs or rocks. Preston (1982) observed individuals resting on the bottom of a shallow, ice-covered pond in British Columbia. Painted Turtles will frequently move during the winter. Moriarty and Lewis (unpublished data) have recorded movement of more than 100 meters (300 feet) in one week during January in a Ramsey County lake.

LIFE HISTORY

The cold-tolerant Painted Turtle can be seen basking on logs as early as mid-April on warm, sunny days. They remain active

Normal Painted Turtle plastron.
Photograph by Barney Oldfield.

until late October. After spending a night sleeping underwater, these diurnal reptiles alternate periods of basking, which usually last two hours, with periods of foraging for food (Ernst and Lovich 2009). They frequently share basking sites with other turtles and can often be seen stacked two or three high. Turtle densities can vary from less than 10 to more than 100 turtles per acre in central Minnesota ponds and wetlands (Gamble 2003; Grace et al. 2010).

Adult Painted Turtles are omnivorous and obtain food by foraging in the water along the bottom and through surface vegetation. According to Vogt (1981), approximately 60 percent of their diet is animal matter (snails, crayfish, insects, fish carrion, and tadpoles), while the remainder is plant material (algae, cattails, and duckweeds). Most animal prey are actively pursued, but occasionally the turtles wait in ambush. Hatchling and juvenile Painted Turtles are carnivorous.

Painted Turtles reach sexual maturity at four to six years of age, and there is evidence that maturity is associated with size rather than age (Ernst and Lovich 2009). The majority of courtship and mating takes place in shallow water during May and early June. A male follows a female, overtakes her, and positions

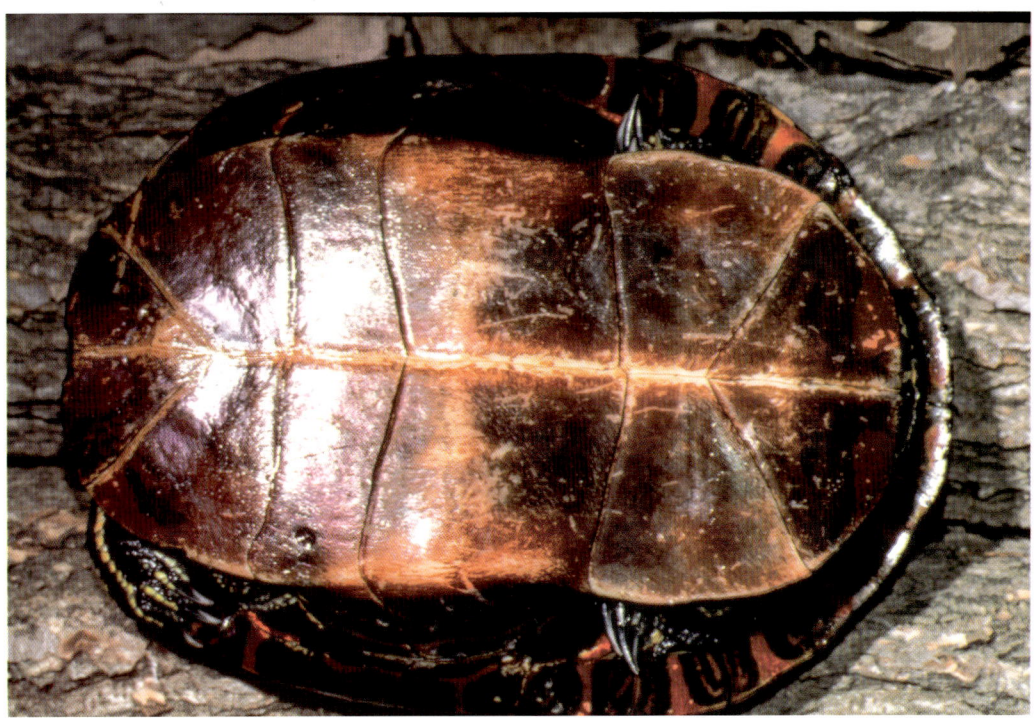

himself in front of and facing her. He uses his long front claws to gently stroke her head and neck. She usually reacts by stroking his front legs with her fore claws. At the conclusion of the courting behavior, she sinks to the bottom. The male mounts her carapace, grasping the edges with his claws, and he curls his tail under hers to achieve copulation (Collins 1982).

Dicephalic Painted Turtle hatchling. Photograph by James E. Gerholdt.

Egg laying begins in late May and may extend well into July, but the majority of the nesting takes place during June in Minnesota. Gravid turtles seek out open areas (south-facing hillsides, sand beaches, railroad rights-of-way, road shoulders, etc.) with loose soil or sand to dig flask-shaped egg chambers with their hind legs. Generally 8 to 9 (range from 2 to 20) white, elliptical eggs with leathery shells comprise a clutch. Average egg size is 3.2 centimeters by 1.9 centimeters (1¼ by ¾ inches) (Vogt 1981). Studies by Moll (1973) in Wisconsin showed that two-thirds of female Painted Turtles in Wisconsin laid two clutches per year.

Hatching begins in late August after 72 to 80 days of development within the egg, but the young turtles frequently stay in the nest until the next spring. Woolverton (1961) studied Painted Turtle nests in northern Minnesota. He found live hatchlings in a nest on October 25 that had been constructed by a female

Nesting Painted Turtle in farm field. Photograph by Minnesota Department of Natural Resources—Carol D. Hall.

turtle on June 25. Hatchlings emerged from the nest the following June after surviving temperatures of -11°C (12.2°F).

While studying Painted Turtle populations, Gamble (2007) found a number of ear abscesses, which produce large lumps on the side of the head. This has lead to incorrect reports of two headed turtles. True two-headed (dicephalic) turtles do occur in Minnesota; such individuals are reported every few years. These turtles normally live only a short time.

Large numbers of Painted Turtle nests are destroyed by raccoons, striped skunks, Virginia opossums, red foxes, and thirteen-lined ground squirrels. Hatchlings and juvenile turtles fall prey to mink, American crows, larger turtles, gartersnakes, Common Watersnakes, American Bullfrogs, and predatory fish. Thousands are killed each year by vehicles.

REMARKS

Of the three recognized subspecies of Painted Turtles, only the Western Painted Turtle (Chrysemys picta bellii) is found in Minnesota (Conant and Collins 1998). A common misnomer for this species is "mud turtle."

Painted Turtles from the Mississippi River backwaters and ponds in the Weaver Dune area of Wabasha County have dark rust-brown plastrons. This coloration is presumably due to absorption of chemicals from the water (Vogt 1981). The original color returns after the turtle sheds its scutes.

Painted Turtles are commercially collected in Minnesota for biological supply companies and the pet trade. The commercial harvest peaked in 1998, when over 60,000 turtles were taken. The annual harvest in 2010 and 2011 averaged 4,000 turtles per year (Larson 2012). Painted Turtles above 5½ inches in length cannot be harvested. Gamble (2003) studied the effect of harvesting Painted Turtles in central Minnesota. He found the harvested lakes had lower populations densities, but all lakes still had a diversity of sizes and sexes.

Longevity can exceed 50 years in the wild (Congdon et al. 2003).

Blanding's Turtle

Emydoidea blandingii

DESCRIPTION

A bright-yellow chin and a helmet-shaped profile are characteristic of this marsh inhabitant. The Blanding's Turtle is a medium
to large turtle with a black or dark-blue, dome-shaped carapace
with muted yellow spots and bars. The large, yellow plastron is
hinged across the anterior third, enabling individuals to pull the
front edge of the plastron firmly up against the carapace to provide additional protection. Large, dark-brown to black blotches
are present on the lateral edge of each plastral scute. The head and
appendages are dark brown or blue gray with small dots of light
brown or yellow. A distinctive field mark is the bright-yellow chin
and throat. This species of turtle has an unusually long neck and
a large rounded head with protruding eyes. A notch in the upper
jaw creates the impression that the turtle is "smiling."

Mature males are significantly larger than females. Males
also have longer tails than females, their cloacal opening is beyond the edge of the carapace, and their plastron is concave.
Hatchlings have a dark-brown or gray carapace with a moderate

Blanding's Turtle, adult female.
Photograph by Allen Blake
Sheldon.

keel. Their plastron has a large, central dark blotch, and the hinge is undeveloped. A hatchling's tail is proportionately longer than an adult's tail. Juvenile Blanding's Turtles are similar in coloration to adults, but the light spots and markings are more conspicuous.

The average adult Blanding's Turtle ranges from 12.5 to 18 centimeters (5 to 7 inches) in straight-line upper-shell length (Conant and Collins 1998). The largest recorded specimen from Minnesota was from Wright County with a carapace length of 28.4 centimeters (11⅛ inches; Blasus, Gerholdt, and Crawford 2004).

DISTRIBUTION

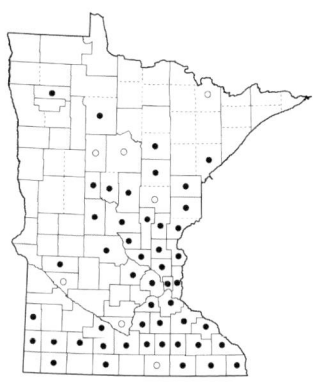

The Blanding's Turtle is known from southern Ontario and the Great Lakes states westward to western Nebraska. It is found as far south as central Illinois. Scattered populations are located in several New England states. Excluding the Arrowhead region, the Blanding's Turtle has a substantial range within Minnesota. It is found in areas of suitable habitat across much of the state, but concentrated populations are known from only a few localities.

HABITAT

Blanding's Turtles favor open areas. Shallow, slow-moving water, mud bottoms, and abundant aquatic vegetation are preferred. Extensive marshes bordering rivers provide excellent habitat (Vogt 1981). In Minnesota, Blanding's Turtles are primarily marsh and pond inhabitants and are frequently found in association with Snapping Turtles and Painted Turtles. They are also found in bogs in east-central Minnesota and in small stream complexes in the southwest region of the state (Moriarty 1986; Lang 2006). Blanding's Turtles attain protection from freezing temperatures by hibernating at the bottom of marshes and ponds.

LIFE HISTORY

Blanding's Turtles emerge from hibernating and begin basking in early April on warm, sunny days. These cold-tolerant turtles have been observed swimming beneath winter ice (Ernst and Lovich 2009). Blanding's Turtles are chiefly diurnal, and they

Blanding's Turtle plastron.
Photograph by John J. Moriarty.

spend large amounts of time basking on muskrat houses, logs, banks, and dikes. Between periods of basking, they forage for food along the bottom of ponds or marshes.

When it comes to food, Blanding's Turtles are not picky. They eat crayfish, frogs, snails, fish, insects, tadpoles, earthworms, slugs, grubs, and occasionally succulent vegetation and berries. Lagler (1940) reported that 56.6 percent of their diet in Michigan was crustaceans. As a rule, aquatic turtles cannot swallow food out of water; however, the Blanding's Turtle is an exception (Ernst and Lovich 2009).

The Blanding's Turtle is a slow-maturing and long-lived species. A nesting female from Chisago County with carved initials on her plastron was determined to be at least 75 years old (Brecke and Moriarty 1989). Congdon, Dunham, and van Loben Sels (1993) reported that female Blanding's Turtles matured between 14 and 20 years of age in central Michigan.

Courtship in Blanding's Turtles has most frequently been observed during the months of April and May, but courting has been reported in other months. Vogt (1981) described courtship and mating in Wisconsin. The ambitious male swims about seeking a female. When a female is found, he approaches her from behind, positions himself on top of her carapace, and begins biting at her head and front limbs. With his neck fully extended, he swings it from side to side in front of the female's head. When she becomes submissive, copulation takes place.

Nesting occurs during the first two weeks of June in Minnesota. Females are most active late in the afternoon and at dusk. Hundreds can be seen crossing roads, fields, and prairies south of Kellogg in Wabasha County during movements to and from nesting localities at the peak of the season. After moving over 1.6 kilometers (1 mile) and going through the laborious process of depositing the eggs, females may stay on land overnight and return to the marsh the next evening (Hamernick 2001; Lang 2000; Linck and Moriarty 1997).

Close-up of Blanding's Turtle showing the yellow throat. Photograph by James E. Gerholdt.

A nest is dug entirely by the female's hind feet in an open, sandy area. Clutch size is reported to be 6 to 15 eggs; however,

most females deposit 8 to 10 elliptical, white eggs with pliable shells that average 3.8 by 2.4 centimeters (1½ by 1 inches) (Vogt 1981). The largest clutch reported for Minnesota was 26 eggs (M. Linck, pers. comm.). During the growth of the turtle embryo, the egg develops thin calcium plaques that make the eggshell brittle. After an egg development period of approximately two months, hatchlings leave the nest in September with an average carapace length of 3.1 centimeters (1¼ inches). Moriarty and Linck (1995) reported a nest that went 126 days between laying and emergence. Linck and Gillette (2009) reported that hatchling Blanding's Turtles will overwinter in wet depressions near nest sites in Scott County. Often, they will make a long, overland trek to find water.

Blanding's Turtles are mild mannered and do not attempt to bite. If molested or threatened, they simply pull into the shell and wait for danger to pass.

Nests and young of Blanding's Turtles fall victim to raccoons, skunks, and predatory birds. Draining of marshes and swamps destroys the habitat of this inoffensive species. A major threat to adult Blanding's Turtles is the automobile. Despite the placement of turtle-crossing signs along a stretch of highway near the town of Weaver, numerous females are killed during the nesting migration each year. The majority of the killing appears to be

Blanding's Turtle habitat in southwest Minnesota. Photograph by Tom Jessen.

intentional, because turtles can easily be seen, and traffic levels are light on the roadway.

REMARKS

The entire plastron of the Blanding's Turtle may be stained dark reddish brown, which obscures the normal pattern. This staining is evident in populations of turtles found along the lower Mississippi River drainage. Normal coloration returns when the scutes are shed. An adult albino female was found and photographed in Washington County during the summer of 1992.

The Blanding's Turtle was named after William Blanding, an early naturalist from Philadelphia. This turtle is classified as threatened, listed as a Species of Greatest Conservation Need in Minnesota, and is given total legal protection (Minnesota DNR 2006, 2013).

Wood Turtle

Glyptemys insculpta

DESCRIPTION

The Wood Turtle, an attractive turtle with a distinctive, sculptured shell, is rare in Minnesota. Each large scute on its broad, rugged carapace is shaped like a low pyramid formed by a stack of concentric growth rings. The back edge of the upper shell is flattened and serrated. The turtle's carapace ranges from light tan to rich dark brown. Some individuals may have prominent yellow lines on each scute arranged in a sunburst pattern. The hingeless plastron is radiant yellow with large, black blotches located laterally on the belly scutes. The undersides of the marginal scutes are similarly patterned. Dorsally, the extremities are dark brown with occasional yellow flecking. The skin covering the soft body parts near the shell, the lower surface of the neck, and the underside of the legs and tail is yellow. The iris of the

Wood Turtle, adult. Photograph by Minnesota Department of Natural Resources—Carol D. Hall.

Wood Turtle's eye is rich brown with a thin, yellow inner margin. In rare individuals, the entire eye color is yellow (Oldfield and Moriarty 1994).

Adult male Wood Turtles are somewhat larger than females. When a male and female are comparable in size, the head of the male is significantly larger than that of the female. Unlike females, males have a concave plastron, a long, thick tail, and large plate-like scales on their forelegs. Hatchlings have a greenish-gray, unkeeled, circular shell that is nearly flat. The hatchling's tail is almost as long as its carapace. Young turtles lack the yellow coloration of adults.

The average adult Wood Turtle has a carapace length ranging from 14 to 20 centimeters (5½ to 8 inches) (Conant and Collins 1998). The largest known Wood Turtle from Minnesota was a male from St. Louis County with a length of 25 centimeters (9⅞ inches) and a weight of 1,860 grams (4.1 pounds) (R. Buech, pers. comm.).

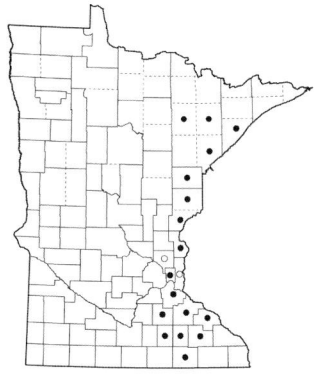

DISTRIBUTION

Wood Turtles are found in the northeastern United States from northern Virginia to Maine, extending westward through parts of southern Canada, Michigan, Wisconsin, and into eastern Minnesota and northeastern Iowa. Records for Wood Turtles come from 16 eastern Minnesota counties. The main populations are in the St. Louis River and Cannon River drainages. Here they are associated with midsize rivers flowing through forested areas. The species is uncommon in the state even in areas of suitable habitat.

HABITAT

The Wood Turtle's primary habitat in Minnesota includes a river with a fairly narrow floodplain and distinct rises to uplands. Much of the floodplain and most of the uplands are wooded (Ewert 1985). Ewert speculates that appropriate habitat extends no farther than 366 meters (400 yards) inland from the home river. In northern Minnesota Wood Turtles with radio transmitters generally stayed within 100 meters (109 yards) of the river (Buech 1995). Wood Turtles share habitat with Snapping Turtles, Painted Turtles, Northern Map Turtles, and Spiny

Softshells. Individual Wood Turtles hibernate in water beneath the ice in bank undercuts, near logjams, or in the sediment without any adjacent structure.

LIFE HISTORY

Wood Turtles become active by late April and begin basking along the banks of the river on warm days. Diurnal in their habits, Wood Turtles are the most terrestrial of Minnesota turtles. Individuals from the western part of the species's range, including Minnesota, remain closer to water than their eastern counterparts (Harding and Bloomer 1979).

According to Ewert (1985), Wood Turtles obtain nearly all of their food on land. This omnivorous turtle generally refuses fish; they eat various succulent forbs, willow leaves, dandelions, strawberries, raspberries, blueberries, earthworms, mushrooms, slugs, and a variety of insects. Occasionally carrion is taken; Farrell and Graham (1991) reported a male eating a dead fish.

The Wood Turtle is a late-maturing species that produces one clutch of eggs per year. Farrell and Graham (1991) reported that the turtles reach sexual maturity at approximately 14 years; however, Buech, Nelson, and Brecke (1990) found that maturity

Hatching twin Wood Turtles.
Photograph by Barney Oldfield.

may be closer to 17 to 18 years of age in more northern locations. Courtship and mating occur primarily in the spring, but fall activity has been observed. The male and female approach each other slowly with necks extended and heads held high. As they come closer, they suddenly lower their heads and swing them from side to side (Carr 1952). The male then climbs astride the female and clamps his claws around the edge of her carapace. He bites at her head and aggressively thumps his plastron against her carapace. Courtship takes place in shallow water, and the pair may remain in copulation for one to two hours (Harding and Bloomer 1979).

Nesting activity occurs on sandbars, riverbanks, open hillsides, and abandoned railroad grades during late May and much of June. Females leave the water in late afternoon to find a nesting site (Oldfield and Moriarty 1994). Sometimes the female digs a shallow body form with her front legs before digging the egg chamber with her rear legs (Farrell and Graham 1991). She deposits a clutch of 4 to 12 (typically 7 to 9) off-white, elliptical eggs with leathery shells averaging 3.2 by 2.3 centimeters (1¼ by ⅞ inches). Once the eggs are deposited, she uses her rear legs to fill the cavity with sand, and then she tamps the nesting site

Wood Turtle nesting beach. Photograph by Minnesota Department of Natural Resources—Carol D. Hall.

Nesting Wood Turtle. Photograph by Minnesota Department of Natural Resources—Carol D. Hall.

with her plastron. The entire procedure may take three hours or longer (Pallas 1960).

Egg development takes 58 to 71 days depending on weather. Young emerge from the nest in late August or September. Hatchling Wood Turtles have a carapace length of 2.8 to 3.8 centimeters (1⅛ to 1½ inches). Wood Turtles are the only turtle known to have generational twining. Moriarty (2006) reported twin embryos in an egg later produced by one of the twin hatchlings.

Large numbers of Wood Turtle nests are destroyed by mammalian predators, such as raccoons, striped skunks, and red foxes. The young fall prey to these predators as well as to large birds and carnivorous fish. Adult turtles probably have few enemies; however, it is not unusual to find wild individuals with amputated toes and tails (B. Oldfield, pers. comm.). When a raccoon encounters a Wood Turtle away from the water, it could conceivably inflict such damage when attempting to eat the turtle. Black bear predation of an adult Wood Turtle has been reported in northern Minnesota (M. Hamady, pers. comm.).

REMARKS

Most Wood Turtles are mild mannered. Newly captured turtles seldom attempt to bite. Typically, they emerge from their protective shell after a short period. A captive was reported to have lived a minimum of 58 years (Oliver 1955). Wood Turtles are capable of a vocalization similar to the sound produced by a muted teakettle. Male turtles are known to emit this sound during courtship (Pope 1939).

In Oldfield and Moriarty (1994) the scientific name of the Wood Turtle was *Clemmys insculpta*. Holman and Fritz (2001) split the genus and moved the Wood Turtle to *Glyptemys*.

The Wood Turtle was once marketed for human food consumption, especially in the eastern part of its range (Harding and Bloomer 1979). In the past, this species has also been extensively collected for the pet trade and biological supply houses. Nearly all states with Wood Turtles have passed legislation to protect this unique and diminishing reptile. Wood Turtles are classified as threatened, listed as a Species of Greatest Conservation Need in Minnesota, and are legally protected by the Minnesota DNR (2006, 2013).

Northern Map Turtle

Graptemys geographica

DESCRIPTION

Highly aquatic, the Northern Map Turtle has a low vertebral keel and a strongly serrated margin on the posterior edge of its carapace. The upper shell is olive green to brown, and there are thin, yellow lines producing a reticulate pattern that is especially evident on the marginal scutes. Recently shed scutes reveal large, dark-brown to black blotches scattered randomly over the carapace. The large plastron is either light yellow or cream colored, and it generally has indistinct, dark smudges along the seams. The olive or brown skin of the head, neck, legs, and tail is embellished with numerous greenish-yellow to bright-yellow lines forming a pattern similar to the contour lines of topographic maps. Marginal undersides are similarly marked with swirling light lines against a dark background. An isolated oval- or triangular-shaped, light-green to yellow spot is located posterior and dorsal to each eye.

Northern Map Turtle, adult female. Photograph by Allen Blake Sheldon.

Males of this species are markedly smaller than females. Males have a stronger keel, pronounced shell markings, a long, thick tail, and long claws on their front feet. The female Northern Map Turtle has a large, massive head and a broad, unmarked upper lip. Hatchling and juvenile Northern Map Turtles have a brightly marked carapace, a pronounced plastral pattern, and a strongly defined keel (Ernst and Lovich 2009).

Adult females have a carapace length of 18 to 27.3 centimeters (7 to 10¾ inches), while males are 9 to 15.9 centimeters (3½ to 6¼ inches) (Conant and Collins 1998).

Northern Map Turtles are easily confused with False and Southern Map Turtles. Positive identification requires a first-hand look at the markings on the head and neck. Northern Map Turtles have an isolated "eyespot" behind and above each eye. Behind the eye of a False Map Turtle is either a spot, blotch, bar, or crescent that is connected to lines of the neck. The Southern Map Turtle has chin spots in addition to the eyebrow. Adult female Northern Map Turtles have proportionally larger heads than do adult female False or Southern Map Turtles.

DISTRIBUTION

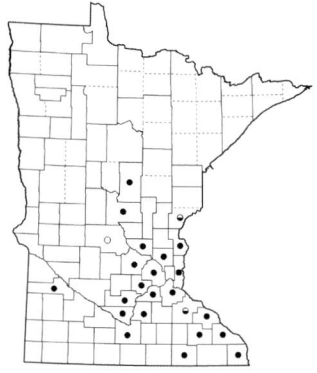

The Northern Map Turtle is a reptile of the north and central river drainages of the eastern United States. Populations are found from Vermont and adjacent Canada west to Minnesota and Kansas and south to Arkansas and central Alabama. New York, New Jersey, Maryland, and Pennsylvania harbor isolated populations. The distribution in Minnesota of this turtle follows the Mississippi, Minnesota, and St. Croix Rivers and their tributaries from the southeastern corner of the state into the central region.

HABITAT

Northern Map Turtles are found in large and midsized rivers. This species requires soft bottoms and ample basking sites surrounded by open water. Backwater sloughs and oxbows within a river's floodplain seem to provide ideal habitat. These turtles hibernate in groups under the ice behind log and rock piles. They also find protection from freezing temperatures in muskrat and beaver channels (Vogt 1981).

Northern Map Turtle plastron.
Photograph by Jeffrey B. LeClere.

LIFE HISTORY

After leaving hibernating sites, Northern Map Turtles begin basking on logs, fallen trees, and mud banks during April and early May. On sunny days they spend considerable time basking, often beginning at eight or nine o'clock in the morning (Evermann and Clark 1916). They are gregarious baskers; as many as 50 to 60 turtles may be stacked 2 and 3 high on a single log. The slightest hint of danger causes them all to plunge into the water. Returning to the water at night, they sleep on the bottom adjacent to submerged logs (Vogt 1981).

Comparison of adult male and female Northern Map Turtles.
Photograph by Barney Oldfield.

Head of Northern Map Turtle.
Photograph by Barney Oldfield.

The large, crushing jaws of female Northern Map Turtles are an adaptation for eating freshwater clams, snails, and crayfish. Males eat aquatic insect larvae and small mollusks. Other food items recorded for Northern Map Turtles include fish carrion, aquatic plants, and water mites (Ernst and Lovich 2009). When foraging for food, these turtles prowl along the river bottom searching through aquatic vegetation.

Spring is the principal mating season, although courting pairs have been observed in the fall (Evermann and Clark 1916). Males actively pursue the larger females during courtship activity. Minnesota females begin nesting in early June, and the egg-laying season extends into early July. It is possible that females deposit two clutches of 10 to 20 eggs each season, although this has not been confirmed in Minnesota. Gravid females leave the water and often travel considerable distance to find an open site to dig a nest (Carr 1952). Northern Map Turtles prefer to nest on overcast days with light rain; however, they also nest during the early morning on sunny days. Females may make several trial excavations before laying eggs. If an intruder happens along while the female is nesting, she pulls in her head and waits for the danger to pass. Eggs are white with leathery shells and are approximately 3.5 centimeters (1⅜ inches) in length and 2.2 centimeters (⅞ inch) in width.

Northern Map Turtle, hatchling. Photograph by Allen Blake Sheldon.

Hatchlings may emerge in early September or hibernate in the nest and leave in May or June of the following year (Vogt 1981). Laboratory experiments demonstrated that the sex of a hatchling is determined by egg development temperature rather than sex chromosomes. Eggs incubated at 30.5°C (86.9°F) produced mostly females, while those incubated at 25°C (77°F) resulted in a high percentage of males (Bull and Vogt 1979). Hatchlings average 3.2 centimeters (1¼ inches) in carapace length (Johnson 1987).

Self-defense for this species includes wariness and prompt use of water as a refuge. Females caught away from water pull into their shell and may attempt to bite if agitated.

Northern Map Turtle nests are destroyed by raccoons, striped skunks, and red foxes; hatchling turtles must avoid a host of birds, fish, and mammals to survive. Many adults from the Mississippi River near Red Wing have amputated toes, missing feet, and damaged shells, indicating predator attack at some time during their life (Oldfield and Moriarty 1994). Collisions with boats and their propellers can damage shells and kill turtles. Environmental pollution and habitat destruction are the greatest threat to their existence.

REMARKS

Northern Map Turtles are protected from commercial harvesting in Minnesota by state statute.

Southern Map Turtle

Graptemys ouachitensis

DESCRIPTION

One of the three map turtles native to Minnesota, the Southern Map Turtle has a boldly marked face. Behind each eye is a large, rectangular or oval, yellow blotch. Occasionally, this postorbital marking is continuous with another prominent spot that occurs just above the angle of the jaw. Located directly below the eye on the mandible is another yellow spot. Additionally, an unpaired spot occurs in the center of the lower jaw, just below the mouth. Most turtles have yellow eyes, but occasionally individuals are found with white irises.

This member of the genus *Graptemys* has a prominent medial keel with blunt projections on the second, third, and fourth vertebrals, and the rear of the carapace is markedly serrated. The oval-shaped upper shell is dark olive, brown, or gray, and each scute has at least one black smudge marking on its posterior edge. Young individuals may display a pattern of yellow or orange interconnected circles, but old adults are often quite dark with little discernable pattern. The light-yellow plastron may be

Southern Map Turtle, adult female. Photograph by Allen Blake Sheldon.

tinted with orange, and there are large, irregular-shaped, black markings along the seams of the scutes. A dark, swirling pattern is noticeable on the bridge, which connects the carapace and the plastron, and on the lower surface of the marginals. Skin coloration is dark green to black with a network of fine, yellow lines on the neck, legs, and tail.

The adult female Southern Map Turtle is considerably larger than the male, a characteristic shared with other members of the genus. Adult females have a carapace length of 12 to 26 centimeters (4¾ to 10¼ inches; Vogt 1981), and males have a carapace length of 9 to 16 centimeters (3½ to 6½ inches). Male turtles have elongated claws on their front feet and a long, well-muscled tail with the cloacal opening beyond the edge of the carapace.

The markings of males and juveniles are bolder than those of adult females, with proportionately larger spots and brighter colors. The vertebral keel of males and juveniles is elevated and has black-tipped projections. Hatchlings are olive green or brown with pale, circular lines on their carapace, and the plastral pattern is significantly less extensive when compared to the plastron of the hatchling False Map Turtle.

Distinguishing between this species and the other two Minnesota map turtles is difficult without seeing head markings. If the turtle is not in hand, careful stalking of basking turtles with a spotting scope is required to make the determination. False Map Turtles have an "eyebrow" mark, and Northern Map Turtles have an isolated postorbital spot, in contrast to the multiple face markings of the Southern Map Turtle.

DISTRIBUTION

Southern Map Turtles range from Minnesota and Wisconsin to Louisiana through the states of the Mississippi River basin. In Minnesota, Southern Map Turtles are found in the Mississippi River from St. Paul to Iowa.

HABITAT

Southern Map Turtles are big-river turtles and are found in areas having relatively strong currents, abundant aquatic vegetation, and ample basking sites away from the shoreline. Vogt (1981) reported that he found them hibernating behind wing dams along

Head of Southern Map Turtle.
Photograph by Barney Oldfield.

the Mississippi River along with Northern Map Turtles and False Map Turtles.

LIFE HISTORY

Seasonal and daily activities of Southern Map Turtles are similar to those of False Map Turtles, which are close relatives. They become active in early May and seek hibernating sites in early October. Diurnal, these turtles spend great amounts of time basking, often stacked two and three high, with other turtles of the same or different species. Basking sites are located on logs and muskrat houses in open water.

Vogt (1981) reported on the feeding habits of Southern Map Turtles. He found that females often feed at the water's surface, consuming vegetation and available insects. Vegetation (pondweed, duckweed, eelgrass, manna grass, arrowhead, and algae) makes up 40 percent of their diet, and aquatic insects make up another 44 percent. Males are primarily carnivorous and forage below the surface.

Courtship probably occurs in the spring and fall, although observations in the field have not been recorded. The male recognizes a female of his species by her head markings and cloacal scents (Vogt 1980). The male swims around to face the female, placing his nose against hers while he drums her ears and neck with his long claws. The number of strokes per drumming bout averages 5.2 and appears to be species specific (Vogt 1983). Copulation occurs after the male mounts the female's carapace.

Southern Map Turtles nest on sandbars and beaches from late May to mid July (Pappas, Congdon, and Capps 2003) within 100 meters (109 yards) of water. Generally, two clutches of 8 to 17 white, elliptical eggs with leathery shells are laid. Eggs average 3.4 centimeters (1⅓ inches) by 2.2 centimeters (⅞ inches). Hatchlings emerge in September from early clutches, and the following spring from late clutches. The average carapace length at hatching is 3 centimeters (1⅛ inches). Nesting sites can have densities of over 100 map turtle nests per acre (Pappas, Congdon, and Capps 2003).

Southern Map Turtle hatching. Photograph by Allen Blake Sheldon.

Sex in this species is determined by the temperature at which the eggs develop and not by sex chromosomes. Eggs incubated at 25°C (77°F) in the laboratory hatched into males, whereas those incubated at 30.5°C (87°F) became females (Bull and Vogt 1979). Like other species of map turtles, the Southern Map Turtle is extremely wary and shy. The turtles rely on the water not only as a home but also as a sanctuary from potential predators.

Southern Map Turtle nests and hatchlings are subject to the same list of predators as other Minnesota turtles. Decapitated adult females have been found on nesting beaches along the Mississippi River (Oldfield and Moriarty 1994). Raccoons, weasels, or mink were probably responsible for these deaths.

REMARKS

The common name used for this turtle in Oldfield and Moriarty (1994) was Ouachita Map Turtle, but it was changed to Southern Map Turtle in Crother (2012) so that a species and subspecies did not have the same common name. The subspecies of Southern Map Turtle in Minnesota is Ouachita Map Turtle (*Graptemys ouachitensis ouachitensis*).

Southern Map Turtles are protected from commercial harvesting in Minnesota by state statute.

False Map Turtle

Graptemys pseudogeographica

DESCRIPTION

The False Map Turtle is a medium-sized turtle of Minnesota rivers with a conspicuous vertebral keel and a single "eyebrow" mark behind each eye. Ground coloration of the carapace ranges from light olive to dark brown, and one or more large, black smudged blotches are present on each scute. Some individuals display a net-like pattern of faint yellow or orange lines on the upper shell. The vertebral keel is formed by black-tipped, blunt projections on the second, third, and fourth scutes down the center of the back. The posterior edge of the oval carapace is serrated. Ventral coloration is light yellow or cream with dark, concentric swirls along the seams and underneath the marginals. The bridge is also marked with dark bars and swirls. Plastral coloration fades to a yellow-brown mottling in adult females. Skin of the appendages, neck, and head is dark olive, dark brown, or black with numerous narrow, yellow lines. Behind each eye is a distinct yellow eyebrow, which may occasionally be divided.

False Map Turtle, adult female. Photograph by Allen Blake Sheldon.

The hatchling's carapace is bright olive green with numerous yellow lines and a black-tipped keel. At least 75 percent of the plastron is covered with interconnecting dark-green or black swirls.

In comparison to adult females, the smaller males have a long tail, longer front claws, and a more pronounced vertebral keel. Adult females are 12 to 27.7 centimeters (4¾ to 10⅞ inches) in carapace length, and males are 9 to 15 centimeters (3½ to 5⅞ inches; Vogt 1993).

Making positive identification of map turtle species in Minnesota requires a close inspection of head and neck markings. With basking turtles, patient stalking with binoculars or spotting scope is required to make the determination. Refer to the species accounts for the Northern Map Turtle and the Southern Map Turtle for an explanation of head and neck markings.

DISTRIBUTION

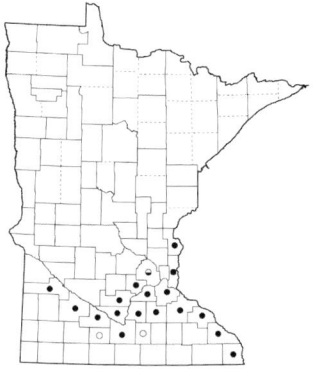

False Map Turtles are found along the Mississippi River and its major tributaries through the heartland of the United States, from extreme southern North Dakota, Minnesota, and Wisconsin to Louisiana and Texas. Minnesota records of False Map Turtles are from counties bordering the St. Croix, Minnesota, and Mississippi Rivers in the southeastern and south-central regions of the state.

HABITAT

In Minnesota, the False Map Turtle is strictly a turtle of the big rivers. This species is found in areas with relatively strong currents, soft bottoms, abundant aquatic vegetation, and ample basking sites away from the shoreline. According to Vogt (1980), these turtles hibernate on the downstream side of logs and rocks on the bottom of rivers. He found them piled up, with other map turtle species, behind wing dams in the Mississippi River.

LIFE HISTORY

False Map Turtles leave hibernating sites in late April and remain active until October. The turtles spend tremendous amounts of time basking with other turtles. Basking sites are

Head of False Map Turtle.
Photograph by Barney Oldfield.

located on logs, stumps, and muskrat houses. At night they sleep on submerged logs or the river bottom.

False Map Turtles eat while swimming, feeding anywhere from the bottom to the surface; 41.5 percent of the turtle's diet is composed of vegetation; 22.5 percent, insects; 16 percent, mollusks; 10 percent, fish; and 10 percent, unidentified matter (Vogt 1981). Lacking the broad, crushing jaws of female Northern Map Turtles, this species feeds less often on clams and snails.

False Map Turtles court and mate in the spring and fall. Swimming around to face the female, the male drums her head and neck with his long front claws (Vogt 1981). Many females double clutch (two egg clutches per season), laying eggs in late May, June, or early July (Pappas, Congdon, and Capps 2003) on open sandbars and beaches near water. The female uses her hind feet to dig a nest about 14 centimeters (5½ inches) deep. Clutch size is 12 to 16 eggs but can range up to 22. The elliptical eggs have pliable shells and average 3.4 by 2.2 centimeters (1⅜ by ⅞ inches). After covering the nest and packing the soil, the female returns to the water. Egg development requires 60 to 75 days, and hatchlings emerge in late summer to early fall or the following spring. Hatchling False Map Turtles are circular shaped and average 3 centimeters (1⅛ inches) in carapace length.

The sex of the False Map Turtle is determined by temperature.

Eggs maintained at 25°C (77°F) in the laboratory produced predominantly males, while eggs maintained at 35°C (95°F) produced females (Vogt 1980).

False Map Turtles are no different than other map turtles in temperament. They are extremely wary and difficult to approach when basking. They rely on water not only as a home but also as a refuge from predators.

REMARKS

The subspecies found in Minnesota is the Northern False Map Turtle *(Graptemys pseudogeographica pseudogeographica)*. The oldest known False Map Turtle lived a minimum of 32½ years in captivity (Bowler 1977).

False Map Turtles are protected from commercial harvesting in Minnesota by state statute.

False Map Turtle, adult male, showing large tail. Photograph by Allen Blake Sheldon.

Pond Slider

Trachemys scripta

DESCRIPTION

The Pond Slider is a medium-sized turtle that has been introduced into Minnesota. The oval, unkeeled carapace is olive tan to green with yellow striping. Dark blotching forms on the scutes in old turtles. The plastron is ivory to yellow with variable dark articulations on the belly scutes. The skin is green to olive with yellow striping. There is a red or yellow head stripe in the areas of the ears.

Pond Sliders average 12.5 to 20.3 centimeters (5 to 8 inches; Conant and Collins 1998) in carapace length. Large specimens can reach 28 centimeters (11 inches).

Pond Slider, adult. Photograph by Allen Blake Sheldon.

DISTRIBUTION

In their native range, Pond Sliders are found from southeast Iowa south to the Gulf Coast and east to the Atlantic. They have been introduced to many other states and can now be found north into Connecticut and in the Pacific Northwest. In Minnesota, Pond Sliders have been found overwintering in Olmsted and Waseca Counties. They have also been observed in other counties, but they have not survived the winter.

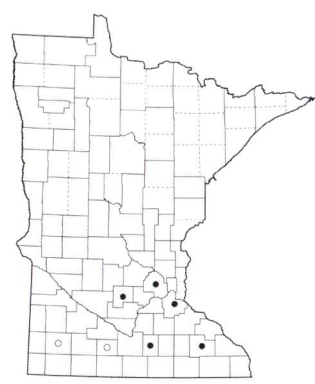

HABITAT

Pond Sliders are found in a variety of habitats, from small wetlands to large rivers. They prefer areas with adequate vegetation and basking sites that are away from the water currents. They use sandy beaches adjacent to the shoreline for nesting habitat.

Pond Slider, hatchling.
Photograph by James E.
Gerholdt.

LIFE HISTORY

Pond Sliders are active from early April to October when the water temperature is over 10°C (50°F; Harding 1997). They are very active baskers and can be found on rocks and logs along the shore.

Pond Sliders are omnivorous and eat a variety of aquatic insects, snails, crustaceans, and aquatic plants (Ernst and Lovich 2009). Juvenile sliders are more carnivorous than adults.

Pond Sliders are a fast-growing turtle; males reach maturity in three to five years, while females mature in six to eight years (Harding 1997). Mating takes place in the spring and fall. Mating courtship is similar to that of Painted Turtles; males use their front claws to caress the female's head.

Females nest from late May to July. They can lay clutches of up to 30 eggs but average 10 eggs (Ernst and Lovich 2009). Pond Slider nests are normally not successful in northern introductions, but nests have been successful around New York City (Gibbs et al. 2007). In Iowa, the closest natural population, Pond Slider nests hatch in late August or September, but hatchlings do not emerge until the following spring (LeClere 2013).

REMARKS

The subspecies found in Minnesota is the Red-eared Slider (*Trachemys scripta elegans*). The red ear stripe gives the subspecies found in Minnesota its common name of Red-eared Slider. This is a very common pet store turtle. They have been sold worldwide by the tens of millions over the years and have become a problem in many states and countries (Kraus 2009).

Pet turtles should never be released into the wild. Besides competing with native populations, they can also spread diseases.

Family Kinosternidae— Musk Turtles

The family Kinosternidae has 25 species in four genera. This family is restricted to North and Central America. These are small turtles that average 10 to 15 centimeters (4 to 6 inches) in length with hinged plastrons. The Eastern Musk Turtle is the only representative of this family in Minnesota.

Eastern Musk Turtle
Sternotherus odoratus

DESCRIPTION

The Eastern Musk Turtle is the smallest turtle in Minnesota. It has a smooth, high-domed carapace, which is uniformly dark brown or black. A layer of algae normally covers the carapace, giving the shell a fuzzy appearance. The cream-colored plastron is reduced in size and has a double hinge. A distinguishing feature of this turtle is the fleshy barbels on the chin and neck. The skin is black or dark brown with two light stripes on each side of the head.

Adult Eastern Musk Turtles are 5 to 12 centimeters (2 to 5 inches) in length. Hatchlings are very small, being no larger than a quarter.

DISTRIBUTION

Eastern Musk Turtles are found throughout the eastern United States and west into Texas.

Eastern Musk Turtles probably occur in the backwaters of

Eastern Musk Turtle, adult. Photograph by Allen Blake Sheldon.

the Mississippi River and tributaries in the southeastern corner of the state. Two specimens were caught in the Zumbro River drainage near Rochester, Minnesota (LeClere 2010). A Trempealeau, Wisconsin, specimen was caught on the shore of the main channel of the Mississippi, in sight of the Winona County, Minnesota, border (G. Casper, pers. comm.). Eastern Musk Turtles are commonly found across the border in the lower Wisconsin River in Wisconsin.

HABITAT

Eastern Musk Turtles are found in rivers, sloughs, lakes, and ponds with soft bottoms. They prefer areas with abundant submergent vegetation. They avoid hard-bottom ponds and lakes, fast-moving rivers, and ephemeral ponds (Vogt 1981).

LIFE HISTORY

Eastern Musk Turtles are a secretive turtle. They are infrequent baskers. When they do bask, they are known to climb up to 2 meters (6 feet) above the water on logs and leaning trees (Conant and Collins 1998).

Eastern Musk Turtles are nocturnal during the warmer part of the summer (Harding 1997). They are mainly carnivorous, eating aquatic insects, snails, fish, and tadpoles. Aquatic plants are occasionally eaten.

Eastern Musk Turtles nest in late May to July. They dig shallow nests in sand but will also use clumps of vegetation and muskrat houses (Ernst and Lovich 2009). Their clutch size averages three to five elliptical eggs that are 2.5 centimeters (1 inch) in length. The eggs have a hard, brittle shell.

Eastern Musk Turtle swimming. Photograph by Tony Gamble.

Eastern Musk Turtle, hatchling. Photograph by Allen Blake Sheldon.

REMARKS

The distribution of this turtle is still mostly unknown in Minnesota. The Rochester, Minnesota, records have been questioned because some think they may be released captives (LeClere 2010). However, Eastern Musk Turtles are not a common pet store turtle, and the chances of catching two released turtles at the same location would be extremely rare. This issue is one reason additional surveys aimed at Eastern Musk Turtles are needed in the southeastern part of the state.

A previous common name for the turtle is "Stinkpot" because of the musky odor it emits.

Family Trionychidae—Softshells

Softshells comprise a relatively small family of turtles with only 30 species in 13 genera. They are found across Asia, Africa, and North America. Softshells have a distinctive flattened shape, a long, snorkel-like nose, and expansive webbing on their feet. Leathery skin covers their carapace and plastron. Softshell turtles are highly aquatic and are fast, agile swimmers. They are also capable of surprising bursts of speed on land.

Of the three species of softshell turtles found in the continental United States, two, the Smooth Softshell and the Spiny Softshell, are found in Minnesota. Both species are at home in major river systems and are occasionally found in large ponds and reservoirs. They are frequently observed on riverbanks and on downed trees at the water's edge.

Smooth Softshell

Apalone mutica

DESCRIPTION

The Smooth Softshell is a medium-sized turtle with delicate-appearing features and a pancake-shaped shell. The completely smooth, keelless carapace is leather-like and flexible, lacking the hard, plate-like scutes common to the majority of turtles. The head has a long, tubular snout, and the feet are paddle-like with extensive webbing. Adult males and juveniles have a light-brown or grayish-tan carapace with numerous scattered, small, dark-brown dots or dashes. Carapace color of adult females is tan or brown with irregular, dark-brown blotches and patches. The plastron of the Smooth Softshell is unmarked white or light gray, and the underlying bones are visible as darker areas of gray. Dorsal coloration of the limbs, tail, head, and neck is similar to the carapace. A black-bordered cream or orange line extends from each eye back onto the turtle's neck. The carapace of hatchlings is light brown with a scattering of small, dark dots and dashes.

Smooth Softshell, adult male. Photograph by Allen Blake Sheldon.

Males have thick tails with the cloacal opening well beyond the edge of the carapace near the tail tip. Females have a narrower tail with the cloacal opening close to the carapace edge. Males are smaller, with a carapace length of 11.5 to 17.8 centimeters (4½ to 7 inches), while females range from 16.5 to 35.6 centimeters (6½ to 14 inches) (Conant and Collins 1998).

The Spiny Softshell is the only turtle species in Minnesota that is easily confused with the Smooth Softshell. Ordinarily, a positive identification is possible after the turtle is in hand. In contrast to the Smooth Softshell, the carapace of the Spiny Softshell may feel like sandpaper. Spines and bumps typically are present along the front edge of the Spiny Softshell's carapace; the front edge of the Smooth Softshell carapace is always smooth. Spiny Softshells have a lateral projection on the nasal septum that extends into the nasal opening. The nasal septum of the Smooth Softshell lacks lateral projections.

DISTRIBUTION

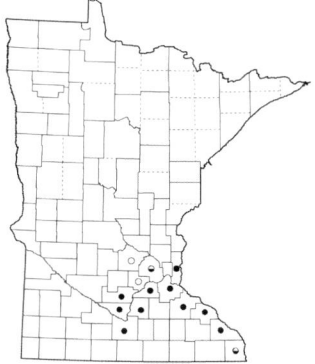

The Smooth Softshell is found across the central United States, ranging from western Pennsylvania to southern North Dakota and south to eastern New Mexico and the Florida panhandle. Its range follows major river systems, including the Mississippi, Missouri, Ohio, Arkansas, and Alabama. The Smooth Softshell is uncommon in Minnesota, where records are confined to the Minnesota, Mississippi, and St. Croix Rivers. Wright County is as far north as they have been found in the state.

HABITAT

In Minnesota, the Smooth Softshell is found only in large rivers with moderate to fast currents. This species prefers sand or mud bottoms. Dense aquatic vegetation and rocky bottoms are avoided. The winter is spent burrowed in the river bottom to avoid freezing temperatures (Collins 1982; Plummer 1977; Plummer and Burnley 1997).

LIFE HISTORY

Smooth Softshells become active in early to mid-May (Vogt 1981). At that time they can be seen basking on sandbars close

Smooth Softshell, hatchling.
Photograph by Barney Oldfield.

to the water's edge. Smooth Softshells spend large amounts of time buried in the sand in shallow water. Periodically, they use their long neck and snout as a snorkel. They are also capable of absorbing oxygen from the water through membranes lining their pharynx and cloaca, thereby allowing them to remain submerged for extended periods of time (Cahn 1937).

The diet of this highly carnivorous turtle includes fish, frogs, tadpoles, mudpuppies, crayfish, aquatic insects, snails, bivalves, and worms (Ernst and Lovich 2009). Plummer and Farrar

Comparison between Smooth *(right)* and Spiny *(left)* Softshells. Photograph by Allen Blake Sheldon.

Smooth Softshell, adult female.
Photograph by Jeffrey B. LeClere.

(1981) found that female turtles in Kansas obtain most of their food from deep water, while the males hunt in shallow water. Much of their food is obtained as they prowl through the water probing with their long neck and nose, but they may also wait in ambush while buried. Highly aquatic, Smooth Softshells are fully capable of chasing down and catching fish.

Breeding most likely occurs in May and June in Minnesota. According to Ernst and Lovich (2009), females mature at seven years of age at a carapace length of 17 to 22 centimeters (6¾ to 8¾ inches), and males become sexually active at a carapace length of 11 to 12.5 centimeters (4¼ to 5 inches). Males actively seek out females by swimming around and investigating other turtles. The receptive female is mounted from behind, and copulation takes place in water.

Nesting occurs in June and early July on sandbars and riverbanks in full sunlight (Vogt 1981). The female uses her hind feet to dig a cavity generally within 18 meters (60 feet) of water. Females are easily frightened by intruders and may abandon the nesting procedure if disturbed. Normally, clutches consist of 15 to 25 eggs, although the range is 4 to 33 (Ernst and Lovich 2009). Eggs resemble Ping-Pong balls and have a thick, hard, white shell and a diameter of 2.0 to 2.3 centimeters (¾ to ⅞ inch). After 8 to 12 weeks of development within the eggs, the

Close-up of Smooth Softshell showing the nostrils. Photograph by Jeffrey B. LeClere.

hatchlings tear through the shell using their front claws. They are equipped with an egg tooth (caruncle) but make less use of it than other turtle species. The sex of Smooth Softshells is determined by sex chromosomes; therefore, the incubation temperature does not affect the sex ratio of hatchlings (Ernst and Lovich 2009).

Hatchling Smooth Softshells are nearly circular in shape and average 4.1 centimeters (1⅝ inches) in carapace length (Ernst and Lovich 2009). They come from the egg with a turned-up snout and a spherical-shaped shell; after a short period, their shell flattens out. The caruncle falls off in 7 to 10 days.

Wariness and speed comprise the basic self-defense plan of Smooth Softshells. They dive in the water and bury themselves in the river bottom at the slightest hint of danger. If surprised on land, they can outrun a human on level terrain (Ernst and Lovich 2009). Adult turtles have few natural enemies, but hatchlings succumb to large fish, turtles, snakes, wading birds, and small mammals. Nests are excavated and the eggs are eaten by skunks, raccoons, and crows.

Plastron of Smooth Softshell. Photograph by John J. Moriarty.

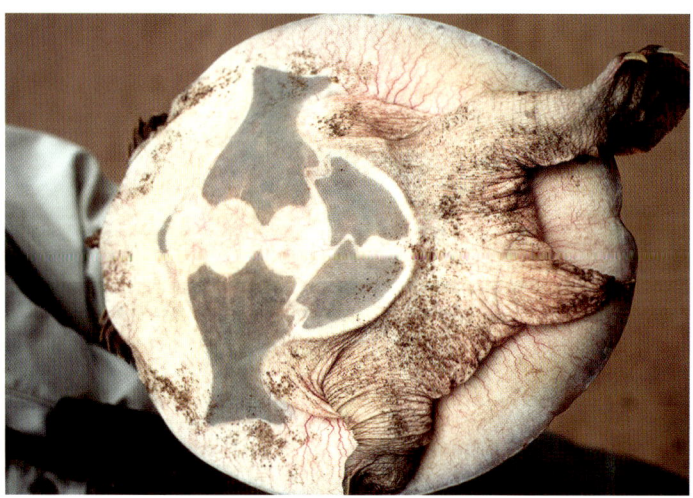

REMARKS

The taxonomic classification of softshell turtles was changed in 1987, and the scientific name of the Smooth Softshell was changed from *Trionyx muticus* to *Apalone mutica* (Meylan 1987). Breckenridge (1944) used the name Brown Soft-shelled Turtle *(Amyda mutica)* to describe this species. According to Conant and Collins (1998), the subspecies of Smooth Softshell found in Minnesota is the Midland Smooth Softshell *(Apalone mutica mutica)*.

Even though Smooth Softshells may feed on small game fish, it is doubtful that this species does any significant damage to fish populations (Ernst and Lovich 2009). Smooth Softshells are timid and rarely scratch or bite when handled, whereas Spiny Softshells tend to be much more aggressive. This is one of the quickest ways to tell the two species apart once they are in hand.

Smooth Softshells are currently listed as special concern in Minnesota and are classified as a Species of Greatest Conservation Need by the Minnesota DNR (2006, 2013).

Spiny Softshell
Apalone spinifera

DESCRIPTION

The Spiny Softshell is a medium to large turtle with a flattened profile, pointed snout, and a leathery shell lacking bony plates. The front edge of the oval-shaped carapace typically has numerous spines or projections in mature turtles. Dorsal coloration of the shell and appendages ranges from olive green to tan with dark-brown or black markings. The carapace of males and immature females is adorned with prominent ocelli (eyelike spots) and a dark line running along the shell margin. Adult females have blotches of light and dark creating a camouflage pattern that makes the ocelli and margin line less obvious. The upper surfaces of the legs and neck are decorated with numerous black blotches and bold lines. The feet have extensive webbing between the toes. A yellow stripe with a black border begins at the nostril, divides into two stripes on top of the snout, and continues through the eyes and down the neck on either side. The nasal

Spiny Softshell, adult male.
Photograph by Barney Oldfield.

septum has a ridge projecting laterally into both nasal openings. Typically, the hingeless white plastron shows dark underlying bones. Juvenile turtles resemble adult males in shape and coloration, but their carapace remains smooth until they mature.

Besides the striking difference in dorsal coloration, adult males are significantly smaller than females. Males are 12.5 to 23.5 centimeters (5 to 9¼ inches) in carapace length, while females are 18 to 45.7 centimeters (7 to 18 inches) (Conant and Collins 1998). The largest Spiny Softshell reported from Minnesota was a female with a carapace length of 49 centimeters (19¼ inches; Pappas and Congdon 2002; Moriarty 2004). Males have longer, thicker tails than females with the cloacal opening very close to the tail tip. The dorsal surface of the male's carapace has a rough sandpaper texture.

In contrast to the Spiny Softshell, the adult Smooth Softshell's carapace is smooth along the front edge. Additionally, the nasal septum of the Spiny Softshell has a transverse ridge (not found in the Smooth Softshell) that projects into the nasal openings.

DISTRIBUTION

The Spiny Softshell ranges from western New York to central Minnesota and then south to Texas and extreme northern Florida. Disjunct populations are found in eastern New York, Vermont, and New Jersey. It is also found in Montana and along the Colorado and Rio Grande Rivers in several southwestern states. In Minnesota, the distribution of the Spiny Softshell follows major river and lake drainages in the central and southern portions of the state. Individuals have also been located in several northern counties in midsized rivers and lakes.

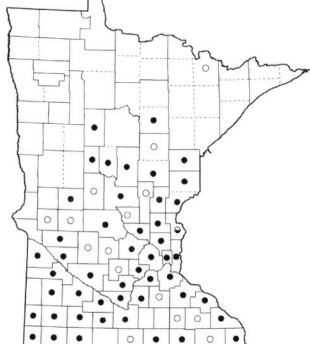

HABITAT

The Spiny Softshell is primarily a river turtle in Minnesota, but it is also found in large lakes and impoundments. Mud or sand bottoms with gravel or sandbars and beaches are important habitat requirements. This species avoids aquatic conditions with rocky bottoms or abundant emergent vegetation. Spiny Softshells hibernate underwater, buried in sand or mud bottoms with heads exposed (Ernst and Lovich 2009). Underwater video of softshells hibernating in the Mississippi show the turtles with their eyes open and aware of the divers (B. Sietman, pers. comm.).

LIFE HISTORY

Spiny Softshells emerge from winter dormancy on warm days during late April and May. This highly aquatic turtle spends considerable time on banks, sandbars, or fallen trees, basking at the river's edge. Time in the water is spent foraging, floating with the current, or buried on the bottom in shallow water. When concealed under the mud or sand, the long neck and nose periodically serve as a snorkel. The vascular lining of the pharynx and cloaca aids in respiration by absorbing dissolved oxygen from the water, allowing the turtle to remain submerged for up to five hours at room temperature, approximately 70°F (24°C) (Dunson 1960).

The carnivorous Spiny Softshell actively searches for food or lies buried waiting to ambush prey. Studies indicate that these turtles feed primarily on invertebrates, including crayfish and

Close-up of Spiny Softshell showing the spines and nostrils. Photograph by Tony Gamble.

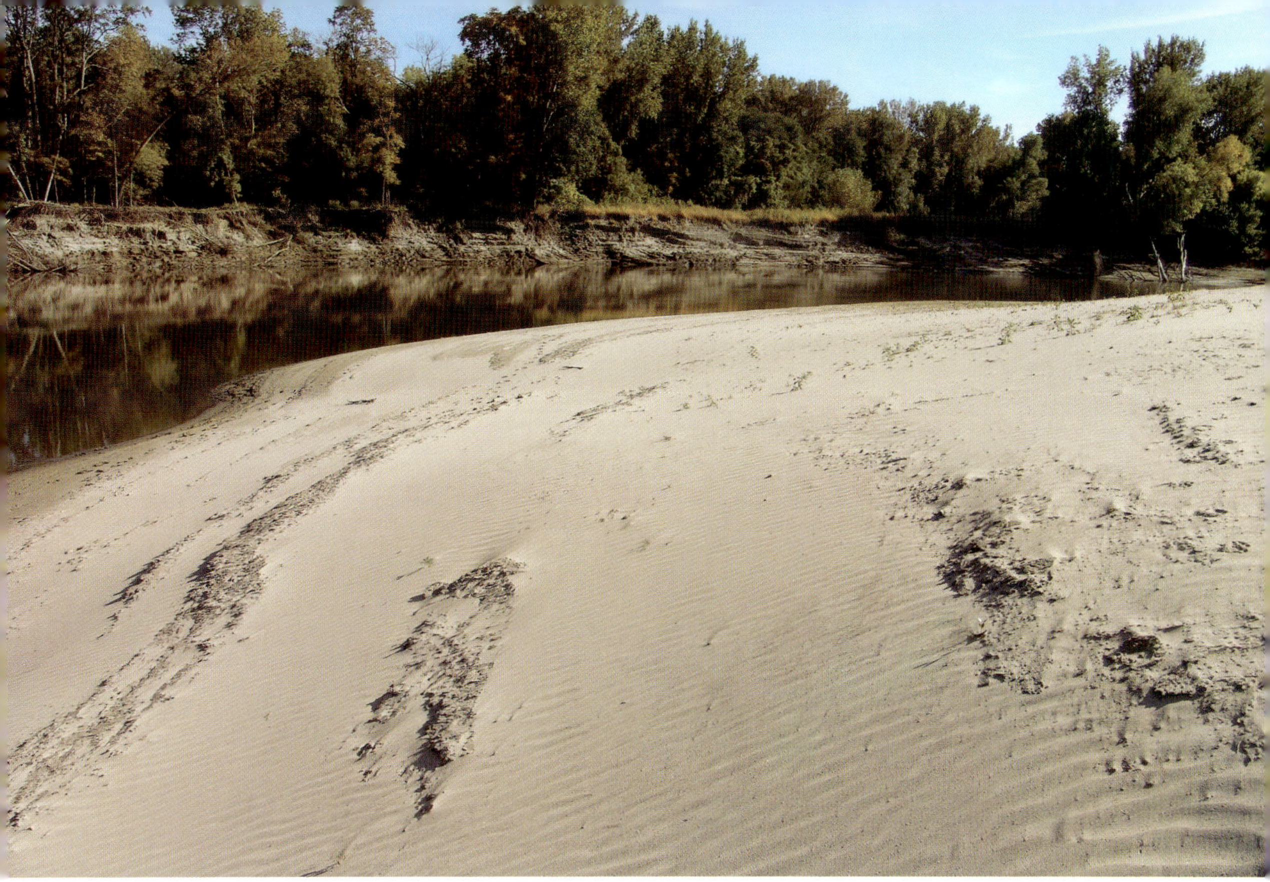

Softshell nesting beach on Minnesota River. Photograph by Minnesota Department of Natural Resources—Carol D. Hall.

aquatic insects. Mollusks, fish, frogs, tadpoles, earthworms, and carrion are also eaten (Ernst and Lovich 2009).

In Minnesota, courtship and mating are primarily springtime events followed by egg laying in June and early July. Female Spiny Softshells select open sandy or gravelly areas close to water to lay their eggs. They excavate an egg cavity to a depth of 10 to 25 centimeters (4 to 10 inches). As they dig, their hind feet spray sand over the top of their carapace. Breckenridge (1960) reported that a female in Minnesota dug a nest in 15 minutes, deposited 17 eggs in 6 minutes, and covered the eggs in 5 minutes. Females often abandon the nesting process and scramble for the water if disturbed.

Overwintering Spiny Softshell with head sticking out of the sand. The turtle was watching the photographer. Photograph by Minnesota Department of Natural Resources—Bernard Sietman.

The spherical, brittle eggs are white and measure 2.8 to 3.2 centimeters (1 to 1¼ inches) in diameter. Clutch size ranges from 4 to 32 eggs, but 15 to 20 is typical. Females may nest two times each season (Ernst and Lovich 2009), but this has not been documented in Minnesota. Hatchlings leave the eggs in late August or early September with a curled nose and a rounded carapace. Several hours after hatching, the carapace flattens

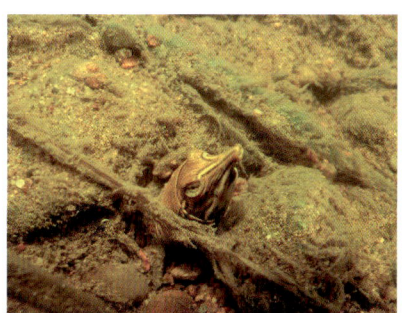

Spiny Softshells hatching.
Photograph by Barney Oldfield.

out to a length of 3.2 to 3.8 centimeters (1¼ to 1½ inches). Hatchlings hibernating in the nest has been reported in Indiana (Minton 1972).

Breckenridge (1955) published growth rate data for Minnesota Spiny Softshells. He reported that a 10-year-old female had a carapace length of 25 centimeters (9⅞ inches); a 15-year-old, 30 centimeters (11¾ inches); a 20-year-old, 33 centimeters (13 inches); a 30-year-old, 38 centimeters (15 inches); and a 53-year-old, 43 centimeters (17 inches). Data for males indicated that a 10-year-old male would be 16 centimeters (6¼ inches), and a 15-year-old, 17 centimeters (6¾ inches).

Spiny Softshells are extremely wary and quickly seek the safety of water if disturbed. When captured they are exceptionally pugnacious. Swift strikes by the long neck, successive attempts to bite with their sharp beak, and rapid raking actions of clawed feet make large individuals difficult to handle. The best control is achieved with a firm grip at the base of the hind legs while holding the turtle well away from the body.

As are those of other turtle species, nests of Spiny Softshells are destroyed by raccoons and striped skunks. Hatchlings fall prey to fish, larger turtles, wading birds, and carnivorous mammals. Large adults have few natural predators, but human activities can have an adverse impact on populations.

REMARKS

Taxonomic changes of softshell turtles by Meylan (1987) resulted in a change of scientific name for the Spiny Softshell from *Trionyx spiniferus* to *Apalone spinifera*. Breckenridge (1944) used the name *Amyda spinifera*. The subspecies of Spiny Softshell found in Minnesota is the Eastern Spiny Softshell (*A. s. spinifera*).

Spiny Softshells are taken for food by turtle trappers in Minnesota. They are a species open for commercial harvest. Commercial trappers harvested fewer than 150 per year in 2010 and 2011 (Larson 2012).

These turtles are frequently referred to as "leatherbacks" or "pancake turtles."

Species of Possible Occurrence

A number of species of amphibians and reptiles are found in adjacent states but have not yet been found in Minnesota. Of these, three species are located within 10 kilometers of the Minnesota border. These species are most likely to be found in the southwest corner of the state, and additional searching will likely eventually document their occurrence.

Three species listed as species of possible occurrence in Oldfield and Moriarty (1994) have been found in the state. They are the Spotted Salamander, Four-toed Salamander, and Eastern Musk Turtle.

The following species accounts for the three species of possible occurrence are abbreviated. They only provide a brief description of the species, their habitat, and potential range within the state. If anyone observes these species in Minnesota, they should try to take a photo and contact the Nongame Wildlife Program (see Resources for contact information) immediately.

RELEASED EXOTIC SPECIES

Over the years a number of exotic (nonnative) species of amphibians and reptiles have been reported from the state. Many species are brought into the state for the pet trade and by accidental "hitchhiking" from other parts of the country. Several of the commonly reported nonnatives are the Eastern Box Turtle *(Terrapene carolina)*, Ornate Box Turtle *(Terrapene ornata)*, and Green Treefrog *(Hyla cinerea)*. When a nonnative species, especially those originating from foreign countries, is encountered in the wild, it should be caught and kept in captivity. They generally will not survive Minnesota winters. If a species does survive and becomes established, it could have negative effects on Minnesota species.

Woodhouse's Toad

Anaxyrus woodhousii

DESCRIPTION

The Woodhouse's Toad is a medium-sized toad (6 to 10 centimeters [2½ to 4 inches]) that has a white belly with no flecking, a light dorsal stripe, and cranial ridges that touch the parotoid glands. It is very similar to the American Toad, but several characteristics distinguish the two species. The American Toad has flecking on the belly, and the cranial crests normally do not touch the parotoid glands. These differences can only be seen with the toad in hand. The breeding calls of the two species are also different. The Woodhouse's Toad has a nasal "waaah" call, while the American Toad produces a high-pitched musical trill (Johnson 1987). Woodhouse's Toads and American Toads are known to hybridize where their ranges overlap (Johnson 1987). This can make identification very difficult.

HABITAT

The Woodhouse's Toad is found in a variety of prairie habitats, including both dry and mesic areas. Breeding takes place in shallow ponds and wetlands.

Woodhouse's Toad. Photograph by Allen Blake Sheldon.

POTENTIAL RANGE

The Woodhouse's Toad is most likely to be found in the southwestern corner of the state in Rock, Nobles, or Pipestone Counties. There is also a possibility of finding this toad along the western border as far north as Clay County. There are currently records in adjacent counties in North Dakota, South Dakota, and Iowa.

Plains Leopard Frog
Lithobates blairi

DESCRIPTION

The Plains Leopard Frog is similar to the Northern Leopard Frog. It tends to be stockier and has a shorter head than the Northern Leopard Frog, but these characteristics are variable. The Plains Leopard Frog has a white spot in the center of the tympanum and a dorsolateral fold that is broken near the lower back. These two characteristics are not found in the Northern Leopard Frog. The Plains Leopard Frog's call is similar to the Northern Leopard Frog's but consists of two to three distinct chucks and does not have a "snore."

HABITAT

Plains Leopard Frogs use grasslands that are associated with permanent bodies of water. In the southwest part of the state they would be found in roadside and drainage ditches and cattle pastures.

POTENTIAL RANGE

The Plains Leopard Frog is most likely to be found in Rock County in the extreme southwestern corner of Minnesota. There is a record from Lincoln County, South Dakota, near the Minnesota border. There is a slight possibility that they may be found in adjacent Pipestone and Nobles Counties.

Plains Leopard Frog. Photograph by James E. Gerholdt.

Plains Spadefoot
Spea bombifrons

DESCRIPTION

The Plains Spadefoot is a small toad reaching up to 5 centimeters (2 inches). The dorsal color is grayish with a reticulation of black and brown. The belly is white, sometimes with spots. The Plains Spadefoot differs from the toads of the genus *Anaxyrus* in Minnesota by having large elliptical eyes and a horny spade on the hind foot.

HABITAT

The Plains Spadefoot is found in arid, sandy grasslands. Individuals are highly fossorial and only come to the surface during wet periods. They use temporary ponds for breeding.

POTENTIAL RANGE

The Plains Spadefoot is most likely to be found in Rock County in the extreme southwestern corner of Minnesota. There is a record within 5 miles of the South Dakota–Minnesota border in Minnehaha County. There is a slight possibility that they may be found in adjacent Pipestone and Nobles Counties.

Plains Spadefoot. Photograph by Allen Blake Sheldon.

Glossary

amelanism. A color condition in which black pigment is lacking.

amniote. A term applied to animals (reptiles, birds, and mammals) with three extraembryonic membranes (amnion, chorion, and allantois) during early development.

amplexus. Clasping behavior of amphibians during breeding.

anal plate. The scale just anterior to the cloacal opening in snakes.

anuran. A tailless amphibian (frog or toad).

arboreal. Living in trees.

beak. Hard edge of a turtle's mouth.

boss. A swollen, rounded projection in the center of the snout.

bufotoxin. A poison produced by toads.

carapace. The dorsal shell of turtles.

caruncle. A horny projection on the end of the snout of a hatching reptile used to slit open the egg during hatching.

cline. A gradual change in a species' characteristics across its geographic range.

cloaca. The common chamber where intestinal, urinary, and reproductive tracts empty.

conspecific. Pertaining to the same species.

cornified. Composed of a hard, horny substance such as keratin.

costal. Referring to the area of the body near the ribs.

costal groove. Vertical furrow on the sides of some salamanders.

cranial. Relating to the head.

cranial crests. Ridges behind or between the eyes of some toads.

diploid. Possessing the common two sets of chromosomes.

disjunct. A population isolated or separated from the general range of the species.

diurnal. Active during daylight.

dorsal. Pertaining to the back or upper surface of the body.

dorsolateral fold. A ridge along the side of the back of some anurans.

ecdysis. The process of shedding the skin.

ectotherm. An animal whose body temperature is largely dependent on its environment.

eft. The land-dwelling stage of a newt.

egg tooth. *See* **caruncle**.

estivation. A period of inactivity during a warm season.

eye spots. Markings on the head and neck that resemble eyes.

facial pit. The heat-sensitive organ between the eye and nostril of pit vipers.

femoral. Pertaining to the area of the thigh.

fossorial. Living underground or under surface cover.

glycerol. A chemical in the blood or cells that acts as antifreeze.

gravid. Pregnant; containing developed eggs.

gular. Pertaining to the area of the throat.

hemipenis. One of the paired, extendible copulatory organs of male lizards and snakes.

herpetologist. A person who studies amphibians or reptiles.

herpetology. The study of amphibians and reptiles.

hibernaculum. The place where an amphibian or reptile overwinters.

intergradation. The sharing of features between two or more subspecies, producing an intermediate specimen.

internasal. Positioned between the nasal openings.

Jacobson's organ. A sensory organ in the roof of the mouth that allows snakes to taste.

keel. A narrow, elongated ridge on the carapace of some turtles and the dorsal scales of some snakes.

keeled scale. The scale of a lizard or snake that has a ridge through it.

labial. Pertaining to the area around the edge of the mouth.

larva. The immature form of amphibians, usually possessing gills.

lateral. Relating to the sides of an animal.

marginal. Pertaining to the outer edge of a turtle's carapace.

melanism. A color condition with black as the dominant pigment.

melanophore. A cell containing black pigment.

mental. Pertaining to the area of the chin.

metamorphosis. The process of body transformation that occurs when an amphibian larva changes into an adult.

middorsal. Pertaining to the area in the middle of the back.

morph. The color, shape, structure, or form of an animal used to make its identification.

nasal septum. The membranous structure that separates the nostrils.

neotenic. Having a condition found in some salamanders where the sexually mature adult retains characteristics of the larval stage.

nocturnal. Active during the hours of darkness.

ocelli. Round markings on an animal that may be made up of concentric rings.

oviparous. Egg-laying.

ovoviviparous. Retaining eggs within the female body until hatching; the young are born enveloped in membranes.

parotoid glands. Large glands located behind the eyes of toads.

parthenogenesis. Reproduction without males by the development of an unfertilized egg.

plastron. The lower shell of a turtle.

plates. Large scales of reptiles, especially those of the upper and lower shells of turtles.

postocular. Behind the eye.

prebutton. The structure on the end of the tail of a newborn rattlesnake.

rostral. Pertaining to the area of the snout or nose.

rostrum. Snout or nose.

scute. Plates forming the upper and lower shells of turtles.

septal. Pertaining to a partition or wall.

serrate. Saw-toothed.

smooth scale. A scale that lacks a keel or ridge.

snout–vent length (SVL). Distance from the tip of the snout to the cloacal opening.

spectacle. The clear scale that covers and protects a snake's eye.

spermatophore. A gelatinous structure containing sperm discharged from the cloaca of male salamanders during breeding activities.

subspecies. A geographic race of a species.

sympatric. Occupying the same habitat or have broadly overlapping ranges.

tadpole. The aquatic larva of frogs and toads.

thermoregulation. The process of regulating body temperature, in reptiles and amphibians by relying on the environment.

toepad. The enlarged end of a toe.

tubercle. A bump or projection on the skin.

tympanum. The rounded eardrum on the side of the head.

vent. External opening of the cloaca.

ventral. Pertaining to the lower surface or belly.

Resources

References cited parenthetically in the text of this book can be found in Literature Cited. General references (some of which are also in Literature Cited) are listed here, separated into several groups: Minnesota, Regional, North American, and General. A list of government agencies and herpetological organizations is also provided.

REFERENCES

Minnesota

Breckenridge, W. J. 1944. *Reptiles and Amphibians of Minnesota.* Minneapolis: University of Minnesota Press.

Christoffel, R., J. Edwards, and B. Perry. 2010. *Snakes and Lizards of Minnesota.* St. Paul: Nongame Wildlife Program, Minnesota Department of Natural Resources.

Coffin, B., and L. Pfannmuller. 1988. *Minnesota's Endangered Flora and Fauna.* Minneapolis: University of Minnesota Press.

Karns, D. R. 1986. *Field Herpetology: Methods for the Study of Amphibians and Reptiles in Minnesota.* Bell Museum of Natural History Occasional Paper 18.

Minnesota Department of Natural Resources. 2006. *Tomorrow's Habitat for the Wild and Rare: An Action Plan for Minnesota Wildlife.* St. Paul: Minnesota Department of Natural Resources.

Moriarty, J. J. 2004. *Turtles and Turtle Watching for North Central States.* St. Paul: Nongame Wildlife Program, Minnesota Department of Natural Resources.

Oldfield, B., and J. J. Moriarty. 1994. *Amphibians and Reptiles Native to Minnesota.* Minneapolis: University of Minnesota Press.

Parmelee, J. R., M. G. Knutson, and J. E. Lyons. 2002. *A Field Guide to Amphibian Larvae and Eggs of Minnesota, Wisconsin, and Iowa.* Washington, D.C.: USGS Biological Resources Division ITR—2002–0004.

Sheldon, A. B. 2006. *Amphibians and Reptiles of the Northwoods.* Duluth: Kollath and Stensaas Publishing.

Tekiela, S. 2003. *Reptiles and Amphibians of Minnesota Field Guide.* Cambridge, Minn.: Adventure Publications.

Tester, J. 1994. *Minnesota's Natural Heritage: An Ecological Perspective.* Minneapolis: University of Minnesota Press.

Regional (Upper Midwest)

Harding, J. H. 1997. *Amphibians and Reptiles of the Great Lakes Region.* Ann Arbor: University of Michigan Press.

Johnson, T. R. 2000. *The Amphibians and Reptiles of Missouri.* Jefferson City: Missouri Department of Conservation.

Kiesow, A. M. 2006. *Field Guide to Amphibians and Reptiles of South Dakota.* Pierre: South Dakota Department of Game, Fish, and Parks.

Kingsbury, B. A., and J. Gibson, eds. 2012. *Habitat Management Guidelines for Amphibians and Reptiles of the Midwestern United States.* Partners in Amphibian and Reptile Conservation Technical Publication HMG-1, 2nd ed.

LeClere, J. B. 2013. *A Field Guide to the Amphibians and Reptiles of Iowa.* Rodeo, N.Mex.: ECO Herpetological Publishing.

Phillips, C. A., R. A. Brandon, and E. O. Moll. 1999. *Field Guide to Amphibians and Reptiles of Illinois.* Champaign: Illinois Natural History Survey.

Preston, W. B. 1982. *Amphibians and Reptiles of Manitoba.* Winnipeg: Manitoba Museum of Man and Nature.

Vogt, R. C. 1981. *Natural History of Amphibians and Reptiles of Wisconsin.* Milwaukee: Milwaukee Public Museum.

Wheeler, G. C., and J. Wheeler. 1966. *Amphibians and Reptiles of North Dakota.* Grand Forks: University of North Dakota.

North American

Bartlett, R. D., and P. P. Bartlett. 2006. *Guide and Reference to the Amphibians of Eastern and Central North America.* Gainesville: University Press of Florida.

———. 2006. *Guide and Reference to the Crocodilians, Turtles, and Lizards of Eastern and Central North America.* Gainesville: University Press of Florida.

———. *Guide and Reference to the Snakes of Eastern and Central North America.* Gainesville: University Press of Florida.

Behler, J. L., and F. W. King. 1979. *Audubon Society Field Guide to North American Reptiles and Amphibians.* New York: Alfred Knopf.

Conant, R., and J. T. Collins. 1998. *A Field Guide to Reptiles and Amphibians of Eastern and Central North America.* 3rd ed. Boston: Houghton Mifflin.

Ernst, C. H., and E. M. Ernst. 2003. *Snakes of North America and Canada.* Washington, D.C.: Smithsonian Books.

Ernst, C. H., and J. Lovich. 2009. *Turtles of the United States and Canada.* Baltimore: Johns Hopkins University Press.

Mitchell, J. C., R. E. Jung Brown, and B. Bartholomew. 2008. *Urban Herpetology.* Salt Lake City: Society for Study of Amphibians and Reptiles.

Petranka, J. W. 1998. *Salamanders of the United States and Canada.* Washington, D.C.: Smithsonian Press.

Powell, R., J. T. Collins, and E. D. Hooper Jr. 2012. *Key to the Herpetofauna of the Continental United States and Canada.* Lawrence: University of Kansas Press.

Smith, H. M. 1978. *A Guide to Field Identification: Amphibians of North America.* New York: Golden Press.

Smith, H. M., and E. D. Brodie Jr. 1982. *A Guide to Field Identification: Reptiles of North America.* New York: Golden Press.

Society for Study of Amphibians and Reptiles. 1962–2012. *Catalogue of American Amphibians and Reptiles.* Salt Lake City: Society for Study of Amphibians and Reptiles.

Stebbins, R. C. 2003. *A Field Guide to Western Reptiles and Amphibians.* 3rd ed. Boston: Houghton Mifflin.

Tennant, A., and R. D. Bartlett. 2000. *Snakes of North America: Eastern and Central Regions.* Houston: Gulf Publishing.

General

Crother, B. 2012. *Scientific and Standard English Names of Amphibians and Reptiles of North America North of Mexico, with Comments regarding Confidence in Our Understanding.* SSAR Herpetological Circular 39.

Duellman, W. E., and L. Trueb. 1985. *Amphibian Biology.* New York: McGraw-Hill.

Elliot, L., C. Gerhardt, and C. Davidson. 2009. *The Frogs and Toads of North America.* Boston: Houghton Mifflin Harcourt.

Halliday, T., and K. Adler. 1986. *Encyclopedia of Reptiles and Amphibians.* New York: Facts on File.

Harless, M., and H. Morlock, eds. 1979. *Turtles: Perspectives and Research.* New York: Wiley and Sons.

Lannoo, M. 2006. *Amphibian Declines: The*

Conservation Status of United States Species.
Berkeley and Los Angeles: University of
California Press.

McDiarmid, R. W., and R. Altig. 1999. *Tadpoles:
The Biology of Anuran Larvae.* Chicago:
University of Chicago Press.

McDiarmid, R. W., M. S. Foster, C. Guyer,
J. W. Gibbons, and N. Chernoff. 2012. *Reptile
Diversity: Standard Methods for Inventory
and Monitoring.* Berkeley and Los Angeles:
University of California Press.

Mullin, S. J., and R. A. Seigel. 2009. *Snakes:
Ecology and Conservation.* Ithaca, N.Y.:
Cornell University Press.

Vitt, L. J., and J. P. Caldwell. 2008. *Herpetology.*
3rd ed. New York: Academic Press.

Wells, K. D. 2007. *The Ecology and Biology of
Amphibians.* Chicago: University of Chicago
Press.

ORGANIZATIONS

**American Society of Ichthyologists and
Herpetologists**
http://www.asih.org
P. O. Box 7065
Lawrence, KS 66044
> PUBLICATION: *Copeia*

Bell Museum of Natural History
http://www.bellmuseum.org
10 Church Street Southeast
Minneapolis, MN 55455
> educational programs, scientific collections,
> exhibits

Center of North American Herpetology
http://www.cnah.org
Sternberg Museum of Natural History
Fort Hays State University
3000 Sternberg Drive
Hays, KS 67601
> amphibian and reptile information transfer

Herpetologists' League
http://www.herpetologistsleague.org
810 East Tenth Street
Lawrence, KS 66044
> PUBLICATIONS: *Herpetologica,* herpetological
> monographs

Minnesota Herpetological Society
http://www.mnherpsoc.com
Bell Museum of Natural History
10 Church Street Southeast
Minneapolis, MN 55455
> PUBLICATIONS: monthly newsletter,
> occasional papers, care pamphlets

Minnesota Nongame Wildlife Program
http://www.dnr.state.mn.us/eco/nongame/
index.html
Box 25, DNR Building
500 Lafayette Road
St. Paul, MN 55155-4025
> frog and toad survey, educational
> publications, research grants, data collections

**North American Field Herping Association,
Midwest Chapter**
http://www.nafta.org/midwest-chapter
http://www.naherp.com

**Partners in Amphibian and Reptile
Conservation**
http://www.parcplace.org/
> • PUBLICATIONS: *Habitat Management
>   Guidelines,* educational materials
> • Year of Turtle and Year of Lizard campaigns

Society for the Study of Amphibians and Reptiles
The largest academic herpetological society in North America.
http://www.ssarherps.org
P. O. Box 58517
Salt Lake City, UT 84158
> PUBLICATIONS: *Journal of Herpetology, Herpetological Review, Catalog of American Amphibians and Reptiles, Herpetological Circulars, Contributions to Herpetology*

U.S. Fish and Wildlife Service (Department of the Interior)
http://www.fws.gov
U.S. Fish and Wildlife Service, Midwest Region
5600 American Boulevard West, Suite 900
Bloomington, MN 55437-1458
> numerous programs and resources

Literature Cited

Acker, P. M., K. C. Kruse, and E. B. Krehbiel. 1986. "Aging *Bufo americanus* by Skeleto-chronology." *Journal of Herpetology* 20: 570–74.

Aleksiuk, M. 1976. "Metabolic and Behavioural Adjustments to Temperature Change in the Red-sided Garter Snake *(Thamnophis sirtalis parietalis):* An Integrated Approach." *Journal of Thermal Biology* 1: 153–56.

Anderson, Y. C., and R. J. Baker. 2002. *Minnesota Frog and Toad Calling Survey 1996–2002.* St. Paul: Minnesota Department of Natural Resources.

Andrews, K. M., J. W. Gibbons, and D. M. Jochimsen. 2008. "Ecological Effects of Roads on Amphibians and Reptiles: A Literature Review." In *Urban Herpetology,* ed. J. C. Mitchell, R. E. Jung Brown, and B. Bartholomew, 121–43. Salt Lake City: Society for the Study of Amphibians and Reptiles.

Arnold, S. J. 1976. "Sexual Behavior, Sexual Interference and Sexual Defense in the Salamanders *Ambystoma maculatum, Ambystoma tigrinum* and *Plethodon jordani.*" *Zeitschrift für Tierpsychologie.* 42: 247–300.

Ashley, E. P., A. Kosloski, and S. A. Petrie. 2007. "Incidence of Intentional Vehicle-Reptile Collisions." *Human Dimensions of Wildlife* 12: 137–43.

Baird, S. F., and J. G. Cooper. 1859. "Report upon the Reptiles Collected on the Survey." In *Explorations and Surveys for a Railroad Route from the Mississippi River to the Pacific Ocean.* Vol. 10. U.S. War Department.

Banning, J. L., A. L. Weddle, G. W. Wahl III, M. A. Simon, A. Lauer, R. L. Walters, and R. N. Harris. 2008. "Antifungal Skin Bacteria, Embryonic Survival, and Communal Nesting in Four-toed Salamanders, *Hemidactylium scutatum.*" *Oecologia* 156: 423–29.

Barbour, R. W. 1971. *Amphibians and Reptiles of Kentucky.* Lexington: University Press of Kentucky.

Becker, M. H., and R. N. Harris. 2010. "Cutaneous Bacteria of the Redback Salamander Prevent Morbidity Associated with a Lethal Disease." *PLoS ONE* 5 (6): e10957.

Bee, M. A. 2007. "Selective Phonotaxis by Male Wood Frogs *(Rana sylvatica)* to the Sound of a Chorus." *Behavioral Ecology and Sociobiology* 61: 955–66.

Bee, M. A., and H. C. Gerhardt. 2002. "Individual Voice Recognition in a Territorial Frog *(Rana catesbeiana).*" *Proceedings of the Royal Society of London Series B* 269: 1443–48.

Behler, J. L., and F. W. King. 1979. *Audubon Society Field Guide to North American Reptiles and Amphibians.* New York: Alfred Knopf.

Bellis, E. D. 1957. "An Ecological Study of the Wood Frog, *Rana sylvatica* Le Conte." Ph.D. diss., University of Minnesota.

Berendzen, P. B., T. Gamble, and A. M. Simons. 2003. "The Genetic Status of Northern Cricket Frogs in Minnesota." Final report submitted to the Nongame Research Program. St. Paul: Minnesota Department of Natural Resources.

Bevier, C. R., D. C. Tierney, L. E. Henderson, and H. E. Reid. "Chorus Attendance and Site Fidelity in the Mink Frog, *Rana septentrionalis:* Are Males Territorial?" *Journal of Herpetology* 40: 160–64.

Bishop, S. C. 1926. "Notes on the Habits and Development of the Mudpuppy (*Necturus maculosus* Rafinesque)." New York State Bulletin, no. 268. Albany: University of the State of New York.

———. 1941. *The Salamanders of New York.* New York State Museum Bulletin, no. 364. Albany: University of the State of New York.

Blair, W. F. 1957. "Changes in Vertebrate Populations under Conditions of Drought." *Cold Spring Harbor Symposia on Quantitative Biology* 22: 273–75.

Blanchard, F. N. 1933. "Late Autumn Collections and Hibernating Situations of the Salamander *Hemidactylium scutatum* (Schlegel) in Southern Michigan." *Copeia* 4: 216.

———. 1936. "Eggs and Natural Nests of the Eastern Ringneck Snake, *Diadophis punctatus edwardsii.*" *Papers of the Michigan Academy of Science* 22: 521–37.

Blasus, R. E., J. Gerholdt, and M. Crawford. 2004. "Natural History Notes, *Emydoidea blandingii,* Maximum Size." *Herpetological Review* 35 (1): 54.

Blaustein, A. R., and L. K. Belden. 2005. "Ultraviolet Radiation." In *Amphibian Declines: The Conservation Status of United States Species,* ed. M. Lannoo, 87–88. Berkeley: University of California Press.

Bleakney, S. 1952. "The Amphibians and Reptiles of Nova Scotia." *Canadian Field-Naturalist* 66: 125–29.

Bonin, J., J.-L. DesGranges, C. A. Bishop, J. Rodrigue, A. Gendron, and J. E. Elliott. 1995. "Comparative Study of Contaminants in the Mudpuppy (Amphibia) and the Common Snapping Turtle (Reptilia), St. Lawrence River, Canada." *Archives of Environmental Contamination and Toxicology* 28: 184–94.

Bowler, J. K. 1977. *Longevity of Reptiles and Amphibians in North America Collections.* SSAR Herpetological Circular, no. 6.

Bragg, A. N. 1940. "Observations on the Ecology and Natural History of Anura. I. Habits, Habitat, and Breeding of *Bufo cognatus* Say." *American Naturalist* 74: 322–49.

———. 1960. "Is *Heterodon* Venomous?" *Herpetologica* 16: 121–23.

Bragg, A. N., and M. Brooks. 1958. "Social Behavior in Juveniles of *Bufo cognatus* Say." *Herpetologica* 14: 141–47.

Brandley, M., A. Schmitz, and T. Reeder. 2005. "Partitioned Bayesian Analyses, Partition, Choice, and the Phylogenetic Relationships of Scincid Lizards." *Systematic Biology* 54: 373–90.

Brecke, B., and J. J. Moriarty. 1989. "Natural History Note. Longevity. *Emydoidea blandingii.*" *Herpetological Review* 20: 53.

Breckenridge, W. J. 1941. "The Amphibians and Reptiles of Minnesota with Special Reference to the Black-Banded Skink, *Eumeces septentrionalis* (Baird)." Ph.D. diss., University of Minnesota.

———. 1943. "The Life History of the Black-Banded Skink, *Eumeces septentrionalis* (Baird)." *American Midland Naturalist* 29: 591–606.

———. 1944. *Reptiles and Amphibians of Minnesota.* Minneapolis: University of Minnesota Press.

———. 1955. "Observations on the Life History of the Softshelled Turtle *Trionyx ferox,* with Especial Reference to Growth." *Copeia* 1955: 5–9.

———. 1960. "A Spiny Soft-Shelled Turtle Nest Study." *Herpetologica* 16: 284–85.

Breckenridge, W. J., and J. R. Tester. 1961. "Growth, Local Movements and Hibernation of the Manitoba Toad, *Bufo hemiophrys.*" *Ecology* 42: 637–46.

Brodie, E. D., Jr. 1977. "Salamander Antipredator Postures." *Copeia* 1977: 523–35.

Brown, C. W. 1968. "Additional Observations on the Function of the Nasolabial Grooves of Plethodontid Salamanders." *Copeia* 1968: 728–31.

Brown, W. S. 1982. "Overwintering Body Temperatures of Timber Rattlesnakes *(Crotalus horridus)* in Northeastern New York." *Journal of Herpetology* 16: 145–50.

———. 1991. "Female Reproductive Ecology in a Northern Population of the Timber Rattlesnake, *Crotalus horridus.*" *Herpetologica* 47: 101–5.

———. 1993. *Biology, Status, and Management of the Timber Rattlesnake* (Crotalus horridus). SSAR Herpetological Circular, no. 22.

Buech, R. B. 1995. "The Wood Turtle: Its Life History, Status, and Relationship with Forest Management." Pages 118–27 in *Proceedings of the 1995 NCASI Central Lake States Regional Meeting*, September 13–14, 1995, Rosemont, Ill.

Buech, R. B., M. D. Nelson, and B. J. Brecke. 1990. "Progress Report: Wood Turtle *(Clemmys insculpta)* Habitat Use on the Cloquet River." Report to the Nongame Wildlife Program. St. Paul: Minnesota Department of Natural Resources.

Bull, J. J., and R. C. Vogt. 1979. "Temperature-Dependent Sex Determination in Turtles." *Science* 206: 1186–88.

Burger, J., and R. T. Zappalorti. 1986. "Nest Site Selection by Pine Snakes, *Pitouphis melanoleucus,* in the New Jersey Pine Barrens." *Copeia* 1986: 116–21.

Burkett, R. D. 1984. "An Ecological Study of the Cricket Frog *Acris crepitans.*" In *Vertebrate Ecology and Systematics: A Tribute to Henry S. Fitch,* ed. R. A. Seigel, L. E. Hunt, J. L. Knight, L. Malaret, and N. L. Zuschlag, 89–103. Lawrence: Museum of Natural History, University of Kansas.

Burton, T. M., and G. E. Likens. 1975. "Salamander Populations and Biomass in the Hubbard Brook Experimental Forest, New Hampshire." *Copeia* 1975: 541–46.

Bury, R. B., and J. A. Whelan. 1984. "Ecology and Management of the Bullfrog." U.S. Fish and Wildlife Service, Resource Publication 155.

Cagle, F. R. 1953. "Two New Subspecies of *Graptemys pseudogeographica.*" Occasional Papers of the Museum of Zoology, University of Michigan 546: 1–17.

Cahn, A. R. 1937. *The Turtles of Illinois.* Illinois Biological Monographs. Urbana: University of Illinois.

Caldwell, J. P. 1982. "Disruptive Selection: A Tail Color Polymorphism in *Acris crepitans* in Response to Differential Predation." *Canadian Journal of Zoology* 60: 2818–28.

Caldwell, R. 1975. "Observations on the Winter Activity of the Red-Backed Salamander, *Plethodon cinereus,* in Indiana." *Herpetologica* 31: 21–22.

Campbell, C. 1977. "Some Threatened Frogs and Toads in Ontario." In *Proceedings of the Symposium on Canada's Threatened Species and Habitats,* ed. T. Mosquin and C. Suchal, 130–31. Ottawa: Canadian Nature Federation, Special Publication, Number 6.

Carr, A. F. 1952. *Handbook of Turtles.* Ithaca, N.Y.: Comstock Press.

Carver, J. 1796. *Travels through the Interior Parts of North America.* London: Key and Simpson.

Christiansen, J. L., and R. R. Burken. 1979. "Growth and Maturity of the Snapping Turtle *(Chelydra serpentina)* in Iowa." *Herpetologica* 35: 261–66.

Christoffel, R., J. Edwards, and B. Perry. 2010. *Snakes and Lizards of Minnesota.* St. Paul: Nongame Wildlife Program, Minnesota Department of Natural Resources.

Churchill, T. A., and K. B. Storey. 1996. "Organ Metabolism and Cryoprotectant Synthesis during Freezing in Spring Peepers *Pseudacris crucifer.*" *Copeia* 1996: 517–25.

———. 1991. "Distribution of the Mudpuppy (*Necturus maculosus*) in Minnesota in Relation to Postglacial Events." *Canadian Field-Naturalist* 105: 400–403.

———. 2008. "A Cottonmouth (*Agkistrodon piscivorus*) in Minnesota, and Historical Reports of Other Pit Vipers Unexpected in the Upper Midwest." *Northeastern Naturalist* 15: 461–64.

Coffin, B., and L. Pfannmuller, eds. 1988. *Minnesota's Endangered Flora and Fauna*. Minneapolis: University of Minnesota Press.

Collins, J. T. 1974. *Amphibians and Reptiles in Kansas*. University of Kansas Museum of Natural History Public Education Series 1.

———. 1982. *Amphibians and Reptiles in Kansas*. 2nd ed. University of Kansas Museum of Natural History Public Education Series 8.

Conant, R. 1958. *A Field Guide to Reptiles and Amphibians of the Eastern North America*. Boston: Houghton Mifflin.

Conant, R., and J. T. Collins. 1998. *A Field Guide to Reptiles and Amphibians of Eastern and Central North America*. 3rd ed. Expanded. Boston: Houghton Mifflin.

Congdon, J. D., A. E. Dunham, and R. C. van Loben Sels. 1993. "Delayed Sexual Maturity and Demographics of Blanding's Turtles (*Emydoidea blandingii*): Implications for Conservation and Management of Long-Lived Organisms." *Conservation Biology* 7: 826–33.

Congdon, J. D., R. D. Nagle, D. M. Kinney, R. C. van Loben Sels, T. Quinter, and D. W. Tinkle. 2003. "Testing Hypotheses of Aging in Long-Lived Painted Turtles (*Chrysemys picta*)." *Experimental Gerontology* 38: 765–72.

Conway, C. H., and W. R. Fleming. 1960. "Placental transmission of 22 131 Na and I in *Natrix*." *Copeia* 1960: 53–55.

Cook, F. C. 1964. "Communal Egg-Laying in the Smooth Green Snake." *Herpetologica* 24: 266.

Cook, F. R. 1983. "An Analysis of Toads of the *Bufo Americanus* Group in a Contact Zone in Central Northern North America." Ottawa: National Museums of Canada Publications Natural Sciences 3.

Cope, E. D. 1889. *The Batrachia of North America*. U.S. National Museum Bulletin, no. 34.

———. 1900. *The Crocodilians, Lizards, and Snakes of North America*. Report of the U.S. National Museum. Washington, D.C.: GPO.

Crother, B., ed. 2012. *Scientific and Standard English Names of Amphibians and Reptiles of North America North of Mexico, with Comments Regarding Confidence in Our Understanding*. 7th ed. SSAR Herpetological Circular, no. 39.

Crother, B. I., M. E. White, J. M. Savage, M. E. Echstut, M. R. Graham, and D. W. Gardner. 2011. "A Reevaluation of the Status of the Foxsnakes *Pantherophis gloydi* Conant and *P. vulpinus* Baird and Girard (Lepidosauria)." *ISRN Zoology* 2011. Article ID 436049.

Davis, Andrew K., and John C. Maerz. 2007. "Spot Symmetry Predicts Body Condition in Spotted Salamanders, *Ambystoma maculatum*." *Applied Herpetology* 4: 195–205.

DeGraaf, R. M. 1991. *The Book of the Toad*. Rochester, Vt.: Park Street Press.

DeGraaf, R. M., and D. D. Rudis. 1983. *Amphibians and Reptiles of New England*. Amherst: University of Massachusetts Press.

Dole, J. W. 1965. "Summer Movements of Adult Leopard Frogs, *Rana pipiens* (Schreber), in Northern Michigan." *Ecology* 46: 236–55.

Dorff, C. J. 1995a. "Conservation of Blanding's Turtles (*Emydoidea blandingii*) in East-Central Minnesota: Impacts of Urban Habitat Fragmentation and Wetland Drawdowns." Master's thesis, University of Minnesota.

Dorff, C. 1995b. "Geographic Distribution: *Hemidactylium scutatum*." *Herpetological Review* 26: 150.

————. 1995c. "Geographic Distribution: *Ambystoma laterale* x *A. jeffersonianum*." *Herpetological Review* 26: 150.

Downs, F. L. 1989. "*Ambystoma maculatum* (Shaw), Spotted Salamander." In *Salamanders of Ohio,* ed. R. A. Pfingsten and F. L. Downs, 108–25. Columbus: Ohio Biological Survey Bulletin.

Dunson, W. A. 1960. "Relation of the Rate of Hyoid Movement to Body Weight and Temperature in Diving Softshell Turtles." *Comparative Biochemistry and Physiology* 19: 597–601.

Dymond, J. R., and F. E. J. Fry. 1932. "Notes on the Breeding Habits of the Green Snake (*Liopeltis vernalis*)." *Copeia* 1932: 102.

Elkan, E. 1965. "Myiasis in Australian Frogs." *Annals of Tropical Medicine and Parasitology* 59: 51–54.

Elliot, L., C. Gerhardt, and C. Davidson. 2009. *The Frogs and Toads of North America.* Boston: Houghton Mifflin Harcourt.

Elwell, L., K. Cram, and C. Johnson. 1981. *Ecology of Reptiles and Amphibians in Minnesota: Proceedings of a Symposium, Cass Lake, MN.*

Ernst, C. H. 1974. "Taxonomic Status of the Red-Bellied Snake, *Storeria occipitomaculata*." *Journal of Herpetology* 8: 347–50.

Ernst, C. H., and R. W. Barbour. 1972. *Turtles of the United States.* Lexington: University Press Kentucky.

————. 1989. *Snakes of Eastern North America.* Fairfax, Va.: George Mason University Press.

Ernst, C. H., and J. E. Lovich. 2009. *Turtles of the United States and Canada.* 2nd ed. Baltimore: John Hopkins University Press.

Everman, B. W., and H. W. Clark. 1916. "The Turtles and Bactrachians of the Lake Maxinkuckee Region." *Proceedings of the Indiana Academy of Science* 1916: 472–518.

Ewert, M. A. 1969. "Seasonal Movements of the Toads *Bufo americanus* and *B. cognatus*

in Northwestern Minnesota." Ph.D. diss., University of Minnesota, Minneapolis.

————. 1984. "Assessment of the Current Distribution and Abundance of the Wood Turtle (*Clemmys insculpta*) in Minnesota and along the St. Croix National Scenic Waterway in Wisconsin." First-year report to the Nongame Wildlife Program. St. Paul: Minnesota Department of Natural Resources.

————. 1985. "Assessment of the Current Distribution and Abundance of the Wood Turtle (*Clemmys insculpta*) in Minnesota and along the St. Croix National Scenic Waterway in Wisconsin." Report to the Nongame Wildlife Program. St. Paul: Minnesota Department of Natural Resources.

Farrell, R. F., and T. E. Graham. 1991. "Ecological Notes on the Turtle *Clemmys Insculpta* in Northwestern New Jersey." *Journal of Herpetology* 25: 1–9.

Ferguson, J. H., and R. M. Thornton. 1984. "Oxygen Storage Capacity and Tolerance of Submergence of a Nonaquatic Reptile and an Aquatic Reptile." *Comparative Biochemistry and Physiology* 774: 183–87.

Ferner, J. 2007. *A Review of Marking and Individual Recognition Techniques for Amphibians and Reptiles.* SSAR Herpetological Circular No. 35: 1–72.

Figiel, C. R., Jr., and R. D. Semlitsch. 1991. "Effects of Nonlethal Injury and Habitat Complexity on Predation in Tadpole Populations." *Canadian Journal of Zoology* 69: 830–34.

Fishbeck, D. W. 1968. "A Study of Some Phases in the Ecology of *Rana sylvatica* LeConte." Ph.D. diss., University of Minnesota, Minneapolis.

Fitch, H. S. 1954. "Life History and Ecology of the Five-Lined Skink, *Eumeces fasciatus*." University of Kansas publications, Museum of Natural History 8: 1–156.

———. 1958a. "Natural History of the Six-Lined Racerunner *(Cnemidophorus sexlineatus)."* University Kansas publications, Museum of Natural History 11: 11–62.

———. 1958b. "Home Ranges, Territories, and Seasonal Movements of Vertebrates of the Natural History Reservation." University of Kansas publications, Museum of Natural History 11: 63–326.

———. 1963a. "Natural History of the Racer *Coluber constrictor."* University of Kansas publications, Museum of Natural History 15: 351–468.

———. 1963b. "Natural History of the Black Rat Snake *(Elaphe o. obsoleta)* in Kansas." *Copeia* 1963: 649–58.

———. 1970. "Reproductive Cycles in Lizards and Snakes." University of Kansas Museum of Natural History Miscellaneous Publications 52: 1–247.

Flageole, S., and R. Leclair Jr. 1992. "Étude démographique d'une population de salamanders *(Ambystoma maculatum)* à l'aide de la méthode squeletto-chronologique." *Canadian Journal of Zoology* 70: 740–49.

Fleming, P. L. 1976. "A Study of the Distribution and Ecology of *Rana clamitans* Latreille." Ph.D. diss., University of Minnesota, Minneapolis.

Fox, W. 1956. "Seminal Receptacles of Snakes." *Anatomical Record* 124: 519–40.

Freedman, W., and P. M. Catling. 1979. "Movements of Sympatric Species of Snakes at Amherstburg, Ontario." *Canadian Field-Naturalist* 93: 399–404.

Friedrich, G. W. 1934. "Taxonomy and Distribution of the Fishes, Amphibia, and Reptiles of Central Minnesota." *Proceedings of the Minnesota Academy of Science* 1: 14–15.

Frost, D. R., T. Grant, J. Faivovich, R. H. Bain, A. Haas, C. F. B. Haddad, R. O. de Sá, A. Channing, M. Wilkinson, S. C. Donnellan, C. J. Raxworthy, J. A. Campbell, B. L. Blotto, P. Moler, R. C. Drewes, R. A. Nussbaum, J. D. Lynch, D. M. Green, and W. C. Wheeler. 2006. *The Amphibian Tree of Life.* Bulletin of the American Museum of Natural History No. 297: 1–370.

Gamble, A. B. 2003. "The Commercial Harvest of Painted Turtles in Minnesota." Master's thesis, University of Minnesota, St. Paul.

Gamble, T. 2007. "Incidence of Aural Abscesses in Painted Turtles *(Chrysemys picta)* Populations in Minnesota." *Chelonion Conservation and Biology* 6: 293–95.

Gamble, T., P. B. Berendzen, H. B. Shaffer, D. E. Starkey, and A. M. Simons. 2008. "Species Limits and Phylogeography of North American Cricket Frogs *(Acris:* Hylidae)." *Molecular Phylogenetics and Evolution* 48: 112–25.

Gamble, T., and A. M. Simons. 2004. "Comparison of Harvested and Nonharvested Painted Turtle Populations." *Wildlife Society Bulletin* 32: 1269–77.

Gendron, A. D. 1999. *Status Report on the Mudpuppy* Necturus maculosus *(Rafinesque), in Canada.* COSEWIC report.

Gerhardt, H. C., M. B. Ptacek, L. Barnett, and K. G. Torke. 1994. "Hybridization in the Diploid-Tetraploid Treefrogs *Hyla chrysoscelis* and *Hyla versicolor."* *Copeia* (1): 51–59.

Gerholdt, J., and B. L. Oldfield. 1987. "Life History Notes: *Chelydra serpentina serpentina.* Size." *Herpetological Review* 18: 73.

Gibbons, J. W., and M. E. Dorcas. 2004. *North American Watersnakes: A Natural History.* Norman: University of Oklahoma Press.

Gibbs, J. P., A. R. Breisch, P. K. Ducey, G. Johnson, J. L. Behler, and R. C. Bother. 2007. *The Amphibians and Reptiles of New York State.* New York: Oxford University Press.

Gibson, A. R., and J. B. Falls. 1975. "Evidence for Multiple Insemination in the Common Garter Snake, *Thamnophis sirtalis."* *Canadian Journal of Zoology* 53: 1362–68.

Gill, D. E. 1978. "The Metapopulation Ecology of the Red-spotted Newt, *Notophthalmus viridescens* (Rafinesque)." *Ecological Monographs* 48: 145–66.

Grace, K. J., J. J. Moriarty, E. Curran, and T. L. Lewis. 2010. "Population Size of *Chrysemys picta* in a Metro Area Lake in Minnesota over a Seven Year Period." 2010 Midwest Fish and Wildlife Conference, Minneapolis (abstract).

Gray, R. H. 1983. "Seasonal, Annual and Geographic Variation in Color Morph Frequencies of the Cricket Frog, *Acris crepitans,* in Illinois." *Copeia* 1983: 300–311.

Green, D. M. 1983. "Allozyme Variation through a Clinal Hybrid Zone between the Toads *Bufo americanus* and *Bufo hemiophrys* in Southeastern Manitoba." *Herpetologica* 39: 28–40.

Green, D. M., and C. Pustowka. 1997. "Correlated Morphological and Allozyme Variation in the Hybridizing Toads *Bufo americanus* and *Bufo hemiophrys.*" *Herpetologica* 53: 218–28.

Greer, A. L., M. Berrill, and P. J. Wilson. 2005. "Five Amphibian Mortality Events Associated with *Ranavirus* Infection in South Central Ontario, Canada." *Diseases of Aquatic Organisms* 67: 9–14.

Gregory, P. T. 1977. *Life-History Parameters of the Red-sided Garter Snake* (Thamnophis Sirtalis Parietalis) in *One Extreme Environment, the Interlake Region of Manitoba.* National Museum of Canada Publications on Zoology, no. 13, 1–44.

Gregory, P. T., and K. W. Stewart. 1975. "Long-Distance Dispersal and Feeding Strategy of the Red-sided Garter Snake *(Thamnophis sirtalis parietalis)* in the Interlake of Manitoba." *Canadian Journal of Zoology* 53: 238–45.

Griffith, H., A. Ngo, and R. Murphy. 2000. "A Cladistic Evaluation of the Cosmopolitan Genus *Eumeces* Wiegmann (Reptilia, Squamata, Scincidae)." *Russian Journal of Herpetology* 7: 1–16.

Groves, J. D. 1980. "Mass Predation on a Population of the American Toad, *Bufo americanus.*" *American Midland Naturalist* 103: 202–3.

Guthrie, J. E. 1926. *The Snakes of Iowa.* Agricultural Experiment Station, Bulletin No. 239. Ames: Iowa State College of Agriculture.

Hahn, D. E. 1968. "A Biogeographic Analysis of the Herpetofauna of the San Lius Valley, Colorado." Master's thesis, Louisiana State University, Baton Rouge.

Hall, C. D. 1997. "Minnesota County Biological Survey: Amphibian and Reptile Results." In *Minnesota's Amphibians and Reptiles: Their Conservation and Status,* ed. J. J. Moriarty and D. G. Jones, 58–62. Lanesboro, Minn.: Serpent's Tale Books.

———. 2002. "Geographic Distribution: *Ambystoma maculatum.*" *Herpetological Review* 33: 315.

Hamernick, M. G. 2001. "Home Ranges and Habitat Selection of Blanding's Turtles *(Emydoidea blandingii)* at the Weaver Dunes, Minnesota." Master's thesis, St. Mary's University, Winona, Minn.

Hamilton, W. J., Jr. 1932. "The Food and Feeding Habits of Some Eastern Salamanders." *Copeia* 1932: 83–86.

———. 1934. "The Rate of Growth of the Toad *(Bufo americanus americanus* Holbrook) under Natural Conditions." *Copeia* 1934: 88–90.

———. 1947. "Hibernation of the Lined Snake." *Copeia* 1947: 209–10.

———. 1948. "The Food and Feeding Behavior of the Green Frog, *Rana clamitans* (Latreille), in New York State." *Copeia* 1948: 203–7.

Hammer, D. A. 1969. "Parameters of a Marsh Snapping Turtle Population at LaCreek Refuge, South Dakota." *Journal of Wildlife Management* 33: 995–1005.

Hammerson, G. A. 1982. *Amphibians and Reptiles in Colorado.* Denver: Colorado Division of Wildlife.

———. 1999. *Amphibians and Reptiles in Colorado.* 2nd ed. Colorado Division of Wildlife Publication. Niwot: University Press of Colorado.

Harding, J. H. 1997. *Amphibians and Reptiles of the Great Lakes Region.* Ann Arbor: University of Michigan Press.

Harding, J. H., and T. J. Bloomer. 1979. "The Wood Turtle, *Clemmys insculpta* . . . A Natural History." *Bulletin of the New York Herpetological Society* 15: 9–26.

Harless, M., and H. Morlock. 1979. *Turtles: Perspectives and Research.* New York: Wiley and Sons.

Harper, F. 1947. "A New Cricket Frog *(Acris)* from the Middle Western States." *Proceedings of the Biological Society of Washington* 60: 39–40.

Harris, J. P., Jr. 1959. "The Natural History of *Necturus.*" *Field and Laboratory* 27 (2): 71–77.

Harris, R. N., and D. E. Gill. 1980. "Communal Nesting, Brooding Behavior, and Embryonic Survival of the Four-Toed Salamander *Hemidactylium scutatum.*" *Herpetologica* 36 (2): 141–44.

Hayes, T. B., L. L. Anderson, V. R. Beasley, S. R. de Sollad, T. Iguchie, H. Ingraham, Patrick Kestemon, et al. 2011. "Demasculinization and Feminization of Male Gonads by Atrazine: Consistent Effects across Vertebrate Classes." *Journal of Steroid Biochemistry and Molecular Biology* 127: 64–73.

Hayes, T., K. Haston, M. Tsui, A. Hoang, C. Hadffele, and A. Vonk. 2002. "Feminization of Male Frogs in the Wild." *Nature* 419: 895–96.

Haywood, C. A., and R. W. Harris. 1972. "Fight between Rock Squirrel and Bullsnake." *Texas Journal of Science* 22: 427.

Healy, W. R. 1974. "Population Consequences of Alternative Life Histories in *Notophthalmus v. viridescens.*" *Copeia* 1974 (1): 221–29.

———. 1975. "Breeding and Postlarval Migration of the Red-Spotted Newt, *Notophthalmus v. viridescens,* in Massachusetts." *Ecology* 56: 673–80.

Heatwole, H. 1962. "Environmental Factors Influencing Local Distribution and Activity of the Salamander, *Plethodon cinereus.*" *Ecology* 43: 460–72.

Hedeen, S. E. 1970. "The Ecology and Life History of the Mink Frog, *Rana septentrionalis* Baird." Ph.D. diss., University of Minnesota, Minneapolis.

Hedges, S. B. 1986. "An Electrophorectic Analysis of Holarctic Hylid Frog Evolution." *Systematic Zoology* 35: 1–21.

Helgen, J. 1997. "The Frogs of Granite Falls: Frogs as Biological Indicators." In *Minnesota's Amphibians and Reptiles: Their Conservation and Status,* ed. J. J. Moriarty and D. G. Jones, 55–57. Lanesboro, Minn.: Serpent's Tale Books.

———. 2012. *Peril in the Ponds.* Amherst: University of Massachusetts Press.

Helgen, J., M. Gernes, D. Bowers, J. Burkhart, D. Carlson, K. Gallagher, S. Goldberg, D. Hoppe, M. J. Lannoo, C. Bursey, J. Canfield, and D. Catron. 1998. "Investigation of Abnormal Frogs in Minnesota." In *Midwest Declining Amphibians Conference,* ed. G. S. Casper. Milwaukee, Wis.

Helgen, J. C., M. C. Gernes, S. M. Kersten, J. W. Chirhart, J. T. Canfield, D. Bowers, R. G. McKinnell, and D. M. Hoppe. 2000. "Field Investigation of Malformed Frogs in Minnesota 1993–1997." In *Investigating Amphibian Declines: Proceedings of the 1998 Midwest Declining Amphibians Conference,* ed. H. Kaiser, G. S. Casper, and N. Bernstein, 96–113. Cedar Falls: Iowa Academy of Science.

Helwig, D. D., and M. E. Hora. 1983. "Polychlorinated Biphenyl Mercury and Cadmium Concentrations in Minnesota Snapping Turtles." *Bulletin of Environmental Contamination and Toxicology* 30: 186–90.

Henderson, C. 1979a. "Guide to the Reptiles and Amphibians of Northeast Minnesota, Region 2." St. Paul: Minnesota Department of Natural Resources.

———. 1979b. "Guide to the Reptiles and Amphibians of East Central Minnesota, Region 4E." St. Paul: Minnesota Department of Natural Resources.

———. 1979c. "Guide to the Reptiles and Amphibians of Lower West Central Minnesota, Region 4W." St. Paul: Minnesota Department of Natural Resources.

———. 1980a. "Guide to the Reptiles and Amphibians of Northwest Minnesota, Region 1S." St. Paul: Minnesota Department of Natural Resources.

———. 1980b. "Guide to the Reptiles and Amphibians of Central Minnesota, Region 3W." St. Paul: Minnesota Department of Natural Resources.

———. 1980c. "Guide to the Reptiles and Amphibians of South West Minnesota, Region 4S." St. Paul: Minnesota Department of Natural Resources.

———. 1980d. "Guide to the Reptiles and Amphibians of Southeast Minnesota, Region 5." St. Paul: Minnesota Department of Natural Resources.

———. 1980e. "Guide to the Reptiles and Amphibians of Metropolitan Minnesota, Region 6." St. Paul: Minnesota Department of Natural Resources.

———. 1980f. "Guide to the Reptiles and Amphibians of Northwest Minnesota, Region 1N." St. Paul: Minnesota Department of Natural Resources.

Hennepin, L. 1698 (1903). *A New Discovery*. 2 vols. Chicago: McClurg and Co.

Henrich, T. W. 1968. "Morphological Evidence of Secondary Intergradation between *Bufo hemiophrys* Cope and *Bufo americanus* Holbrook in Eastern South Dakota." *Herpetologica* 24: 1–13.

Hess, Z. J., and Harris, R. N. 2000. "Eggs of *Hemidactylium scutatum* (Caudata: Plethodontidae) Are Unpalatable to Insect Predators." *Copeia* 2: 597–600.

Hillis, D. M., and T. P. Wilcox. 2005. "Phylogeny of the New World True Frogs *(Rana).*" *Molecular Phylogenetics and Evolution* 34: 299–314.

Hoaglund, E. P., and C. E. Smith. 2012. "Lessons Learned: Notes on the Natural History of the Plains Hognose Snake *(Heterodon nasicus)* in Minnesota." *IRCF Reptiles and Amphibians* 19: 163–69.

Holloway, A. K., D. C. Cannatella, H. C. Gerhardt, and D. M. Hillis. 2006. "Polyploids with Different Origins and Ancestors from a Single Sexual Polyploid Species." *American Naturalist* 167: 4 E88–E101.

Holman, J. A. 1998. "Amphibian Recolonization of Midwestern States in the Postglacial Pleistocene." In *Status and Conservation of Midwest Amphibians*, ed. M. J. Lannoo, 9–15. Iowa City: University of Iowa Press.

Holman, J., and U. Fritz. 2001. "A New Emydine Species from the Middle Miocene (Barstovian) of Nebraska, USA with a New Generic Arrangement for the Species of *Clemmys* sensu McDowell (1964) (Reptilia: Testudines: Emyidae)." *Zoologische Abhandlungen-Staatliches Museum für Tierkunde in Dresden* 51: 331–54.

Hoppe, D. M. 1981. "Chorus Frogs and Their Colors." In *Ecology of Reptiles and Amphibians in Minnesota: Proceedings of a Symposium, Cass Lake, MN*, ed. L. Elwell, K. Cram, and C. Johnson.

———. 2000. "History of Minnesota Frog Abnormalities: Do Recent Findings Represent a New Phenomenon?" In *Investigating Amphibian Declines: Proceedings of the 1998 Midwest Declining Amphibians Conference*, ed. H. Kaiser, G. S. Casper, and N. Bernstein, 86–90. Cedar Falls: Iowa Academy of Science.

Hoppe, D. M., and R. G. McKinnell. 1989.

"Report on 1987–1988 Leopard Frog Surveys." Report to the Nongame Wildlife Program. St. Paul: Minnesota Department of Natural Resources.

———. 1991a. "Minnesota's Mutant Leopard Frogs." *Minnesota Volunteer* 54 (319): 56–63.

———. 1991b. "Distribution and Prevalence of Kandiyohi Leopard Frog." Report to the Nongame Wildlife Program. St. Paul: Minnesota Department of Natural Resources.

Hornfeldt, C. S., and D. E. Keyler. 1987. "Review of the Toxicity of Hognose Snakes." In *Occasional Papers of the Minnesota Herpetological Society*, no. 1, 85–90.

Huheey, J. E. 1958. "Some Feeding Habits of the Eastern Hog-Nose Snake." *Herpetologica* 14: 68.

Irwin, J. T., and R. E. Lee Jr. 2003. "Geographic Variation in Energy Storage and Physiological Responses to Freezing in the Gray Treefrogs *Hyla versicolor* and *H. chrysoscelis*." *Journal of Experimental Biology* 206: 2859–67.

Iverson, J. B. 1990. "Nesting and Parental Care in the Mud Turtle, *Kinosternon flavescens*." *Canadian Journal of Zoology* 68 (2): 230–33.

James, J. E. 1966. "Biology of the Soft-Shelled turtle, *Trionyx* sp., of the Upper Mississippi River." Master's thesis, Winona State College, Winona, Minn.

Jaslow, A. P., and R. C. Vogt. 1977. "Identification and Distribution of *Hyla versicolor* and *Hyla chrysoscelis* in Wisconsin." *Herpetologica* 33: 201–5.

Jenks, A. E. 1936. *Pleistocene Man in Minnesota*. Minneapolis: University of Minnesota Press.

Jessen, T. 1991. "A Fox Snake Hibernation Den at the Red Jacket Bridge." Unpublished report to Minnesota Nongame Wildlife Program. St. Paul: Minnesota Department of Natural Resources.

Johnson, B. K., and J. L. Christiansan. 1976. "The Food and Food Habits of Blanchard's Cricket Frog, *Acris crepitans blanchardi*

(Amphibia, Anura, Hylidae), in Iowa." *Journal of Herpetology* 10: 63–74.

Johnson, D. H., S. C. Fowle, and J. A. Jundt. 2000. "North American Reporting Center for Amphibian Malformations." In *Investigating Amphibian Declines: Proceedings of the 1998 Midwest Declining Amphibians Conference*, ed. H. Kaiser, G. S. Casper, and N. Bernstein, 123–27. Cedar Falls: Iowa Academy of Science.

Johnson, J. R., J. H. Knouft, and R. D. Semlitsch. 2007. "Sex and Seasonal Differences in the Spatial Terrestrial Distribution of Gray Treefrog *(Hyla versicolor)* Populations." *Biological Conservation* 140: 250–58.

Johnson, P. T. J., and K. B. Lunde. 2005. "Parasite Infection and Limb Malformations: A Growing Problem in Amphibian Conservation." In *Amphibian Declines: The Conservation Status of United States Species*, ed. M. Lannoo, 124–37. Berkeley: University of California Press.

Johnson, T. R. 2000. *The Amphibians and Reptiles of Missouri*. Jefferson City: Missouri Department of Conservation.

Jones, D. 1981. "Minnesota Herpetological Society: A Progress Report." In *Ecology of Reptiles and Amphibians in Minnesota: Proceedings of a Symposium, Cass Lake, MN*, ed. L. Elwell, K. Cram, and C. Johnson, 19–20.

Jordan, O. R. 1967. "The Occurrence of *Thamnophis sirtalis* and *T. radix* in the Prairie-Forest Ecotone West of Itasca State Park, Minnesota." *Herpetologica* 23: 303–8.

Jordon, R., Jr. 1970. "Death-Feigning in a Captive Red-bellied Snake, *Storeria occipitomaculata* (Storer)." *Herpetologica* 26: 466–68.

Kalb, H. J., and G. R. Zug. 1990. "Age Estimates for a Population of American Toads, *Bufo americanus* (Salientia: Bufonidae), in Northern Virginia." *Brimleyana* 16: 79–86.

Karns, D. R. 1984. "Toxic Bog Water in Northern

Minnesota Peatlands: Ecological and Evolutionary Consequences for Breeding Amphibians." Ph.D. diss., University of Minnesota, Minneapolis.

———. 1986. "Field Herpetology: Methods for the Study of Amphibians and Reptiles in Minnesota." Bell Museum of Natural History Occasional Paper 18.

———. 1992a. "Amphibians and Reptiles in Peatlands." In *The Patterned Peatlands of Minnesota,* ed. H. E. Wright Jr., B. A. Coffin, and N. Aaseng. Minneapolis: University of Minnesota Press.

———. 1992b. "Effects of Acidic Bog Habitats on Amphibian Reproduction in a Northern Minnesota Peatland." *Journal of Herpetology* 26: 401–12.

Kats, L. B., J. W. Petranka, and A. Sih. 1988. "Antipredator Defenses and Persistence of Amphibian Larvae with Fishes." *Ecology* 69: 1865–70.

Keating, W. H. 1825. *Narrative of an Expedition to the Source of the St. Peter's River, Lake Winnepeck, Lake of the Woods.* 2 vol. London: Whittacker.

Kelleher, K. E., and J. R. Tester. 1969. "Homing and Survival in the Manitoba Toad, *Bufo hemiophrys* in Minnesota." *Ecology* 50: 1040–48.

Keyler, D. E. 1983. "Report on Snakebite Fatality." *Minnesota Herpetological Society News* 3 (8): 2.

———. 2005. "Venomous Snakebites: Minnesota and Upper Mississippi River Valley 1982–2002." Minnesota Herpetological Society Occasional Paper Number 7.

Keyler, D. E., and P. A. Cochran. 2005. *Timber Rattlesnake* (Crotalus horridus) *Biology and Conservation in the Upper Mississippi Valley.* Winona, Minn.: St. Mary's Press.

Keyler, D. E., and K. Fuller. 1999. "Survey of Timber Rattlesnake *(Crotalus horridus)* Peripheral Range on Southern Minnesota State Lands (1998)." Final report to the Nongame Wildlife Program. St. Paul: Minnesota Department of Natural Resources.

Keyler, D. E., and B. L. Oldfield. 1992. "Timber Rattlesnake Survey on Southeastern Minnesota State Lands (1990–1991)." Report to the Nongame Wildlife Program. St. Paul: Minnesota Department of Natural Resources.

Keyler, D. E., and J. Wilzbacher. 2002. "Timber Rattlesnake Reproduction at Great River Bluffs State Park (2000–2002), Minnesota." Final Report to the Nongame Wildlife Program. St. Paul: Minnesota Department of Natural Resources.

Kingsbury, B. A., and J. Gibson, eds. 2012. *Habitat Management Guidelines for Amphibians and Reptiles of the Midwestern United States.* Partners in Amphibian and Reptile Conservation Technical Publication HMG-1, 2nd ed.

Kitchell, J., and T. Bergeson. 2010. "Frog and Toad Survey, 2010." Madison: Wisconsin Department of Natural Resources.

Klauber, L. M. 1972. *Rattlesnakes: Their Habits, Life Histories, and Influence on Mankind.* 2 vol. 2nd ed. Berkeley: University California Press.

Kleeberger, S. R., and J. K. Werner. 1982. "Home Range and Homing Behavior of *Plethodon cinereus* in N. Michigan." *Copeia* 1982: 409–15.

Kramer, D. C. 1974. "Home Range of the Western Chorus Frog *Pseudacris triseriata triseriata.*" *Journal of Herpetology* 8: 245–46.

Kraus, F. 2009. *Alien Reptiles and Amphibians: A Scientific Compendium and Analysis.* New York: Springer.

Krupa, J. J. 1986. "Multiple Clutch Production in the Great Plains Toad." *Prairie Naturalist* 18: 151–52.

———. 1994. "Breeding Biology of the Great Plains Toad in Oklahoma." *Journal of Herpetology* 28 (2): 217–24.

Kumpf, K. F. 1934. "The Courtship of *Ambystoma tigrinum." Copeia* 1934: 7–10.

Lagler, K. F. 1940. "A Turtle Loss?" *American Wildlife* 29: 41–44.

Lang, J. W. 1969. "Hibernation and Movements of *Storeria occipitomaculata* in Northern Minnesota." *Journal of Herpetology* 3: 196–97.

———. 1971. "Overwintering of Three Species of Snakes in Northwestern Minnesota." Master's thesis, University of North Dakota, Grand Forks.

———. 1982. "Distribution and Abundance of the Five-lined Skink *Eumeces fasciatus* in Minnesota." Report to the Minnesota Department of Natural Resources, St. Paul.

———. 2000. "Blanding's Turtles, Roads, and Culverts at Weaver Dunes." Report to the Nongame Wildlife Program. St. Paul: Minnesota Department of Natural Resources.

———. 2002. "Blanding's Turtles Studies in Southwest Minnesota." Report to the Nongame Wildlife Program. St. Paul: Minnesota Department of Natural Resources.

———. 2006. "Conservation of Blanding's Turtles in Southwest Minnesota." Report to the Nongame Wildlife Program. St. Paul: Minnesota Department of Natural Resources.

Lang, J. W., D. Karns, D. Wells, M. Nehl, and M. Pappas. 1982. "Status Report on Minnesota's Amphibians and Reptiles." Report to Endangered Species Advisory Committee. St. Paul: Minnesota Department of Natural Resources.

Lannoo, M. J. 1998. Introduction to *Status and Conservation of Midwestern Amphibians,* ed. M. J. Lannoo. Iowa City: University of Iowa Press.

———. 2008. *Malformed Frogs.* Berkeley: University of California Press.

Lannoo, M. J., K. Lang, T. Waltz, and G. S. Phillips. 1994. "An Altered Amphibian Assemblage: Dickinson County, Iowa, 70 Years after Frank Blanchard's Survey." *American Midland Naturalist* 131: 311–19.

Larson, K. A. 2004. "Advertisement Call Complexity in Northern Leopard Frogs, *Rana pipiens." Copeia* 2004: 676–82.

Larson, K. 2009. "Minnesota Frog and Toad Calling Survey: 2009 Survey Results." Report to the Nongame Wildlife Program. St. Paul: Minnesota Department of Natural Resources.

———. 2012. "Minnesota Commercial Turtle Harvest 2011." Report to the Nongame Wildlife Program. St. Paul: Minnesota Department of Natural Resources.

LeClere, J. B. 2010. "Occurrence of Red-eared Slider, *Trachemys scripta elegans,* with a Note Regarding the Common Musk Turtle, *Sternotherus odoratus." Newsletter of the Minnesota Herpetological Society* 30 (2): 6–9.

———. 2011. "A History of MHS Field Surveys." Minnesota Herpetological Society Occasional Paper Number 8.

———. 2013. *A Field Guide to the Amphibians and Reptiles of Iowa.* Rodeo, N.Mex.: ECO Herpetological Publishing.

Legler, J. M. 1955. "Observations on the Sexual Behavior of Captive Turtles. *Llyodia* 18: 95–99.

Lehtinen, R. M., S. M. Galatowitsch, and J. R. Tester. 1999. "Consequences of Habitat Loss and Fragmentation for Wetland Amphibian Assemblages." *Wetlands* 19: 1–12.

Levell, J. P. 1994. "The Search for *Sistrurus* in Minnesota." Unpublished report to the Minnesota Nongame Program. St. Paul: Minnesota Department of Natural Resources.

Linck, M. H. 1988. "*Emydoidea blandingii* Survey within Ramsey Co., MN." Report to the Nongame Wildlife Program. St. Paul: Minnesota Department of Natural Resources.

Linck, M. 2000. "Reduction in Road Mortality

in a Northern Leopard Frog Population." In *Investigating Amphibian Declines: Proceedings of the 1998 Midwest Declining Amphibians Conference,* ed. H. Kaiser, G. S. Casper, and N. Bernstein, 209–11. Cedar Falls: Iowa Academy of Science.

Linck, M., and L. N. Gillette. 2009. "Post-natal Movements and Overwintering Sites of Hatchling Blanding's Turtles, *Emydoidea blandingii*, in East-Central Minnesota." *Herpetological Review* 40: 411–14.

Linck, M., and J. J. Moriarty. 1997. "The Effects of Habitat Fragmentation on Blanding's Turtles in Minnesota." In *Minnesota's Amphibians and Reptiles: Their Conservation and Status,* ed. J. J. Moriarty and D. Jones, 30–37. Lanesboro, Minn.: Serpent's Tales Books.

Livezey, Robert L. 1950. "The Eggs of *Acris gryllus crepitans* Baird." *Herpetologica* 6: 139–40.

Mahaffy, J. 2009. "Historic Extensions of Rattlesnake Ranges Timber Rattlesnake and Massasauga from Five Counties in Southeastern and Southcentral Minnesota." In ASIH meetings July 2009, 22–27. Abstract #400.

Marshall, J. C., Jr., J. W. Manning, and B. A. Kinsbury. 2006. "Movement and Macrohabitat Selection of the Eastern Massasauga in Fen Habitat." *Herpetologica* 62: 141–50.

Martin, W. F., and R. B. Huey. 1971. "The Function of the Epiglottis in Sound Production (Hissing) of *Pituophis melanoleucus*." *Copeia* 1971: 752–54.

Martin, W. H. 1966. "Life History of the Timber Rattlesnake *Crotalus horridus*." Investigator's Annual Report. U.S.D.I. National Park Service. Shenandoah National Park.

Matson, T. O. 1990. "Estimation of Numbers for a Riverine *Necturus* Population before and after TFM Lampricide Exposure." *Kirtlandia* 45: 33–38.

Matthews, Y. 1990. "Progress Report:

Five-lined Skink Electrophorectic Studies in Minnesota." Report to the Nongame Wildlife Program. St. Paul: Minnesota Department of Natural Resources.

McAlister, W. H. 1963. "A Post-breeding Concentration of the Spring Peeper." *Herpetologica* 19: 293.

McCallum, M. L., and S. E. Trauth. 2006. "An Evaluation of the Subspecies *Acris crepitans blanchardi* (Anura, Hylidae)." *Zootaxa* 1104: 1–21.

McComb, W. C., and R. E. Noble. 1981. "Herpetofaunal Use of Natural Tree Cavities and Nest Boxes." *Wildlife Society Bulletin* 9: 261–67.

McGregor, J. H., and W. R. Teska. 1989. "Olfaction as an Orientation Mechanism in Migrating *Ambystoma maculatum*." *Copeia* 1989: 779–81.

McKinistry, D. M. 1978. "Evidence of Toxic Saliva in Some Colubrid Snakes of the United States." *Toxicon* 16: 523–34.

Merrell, D. J. 1965. "The Distribution of the Dominant Burnsi Gene in the Leopard Frog." *Evolution* 19: 69–85.

———. 1970. "Migration and Gene Dispersal in *Rana pipiens*." *American Zoology* 10: 47–52.

———. 1977. "Life History of the Leopard Frog, *Rana pipiens*, in Minnesota." Bell Museum of Natural History Occasional Paper 15.

Meylan, P. A. 1987. "The Phylogenetic Relationships of Soft-Shelled Turtles (Family Trionychidae)." *Bulletin of the American Museum of Natural History* 186: 1–101.

Minnesota DNR. 1998. "The Timber Rattlesnake in Minnesota." St. Paul: Nongame Wildlife Program, Minnesota Department of Natural Resources.

———. 2006. "Tomorrow's Habitat for the Wild and Rare: An Action Plan for Minnesota Wildlife." St. Paul: Minnesota Department of Natural Resources.

———. 2013. "Revised Minnesota List of Endangered, Threatened, and Special

Concern Species." St. Paul: Minnesota Department of Natural Resources.

Minton, S. A., Jr. 1972. *Amphibians and Reptiles of Indiana*. Indiana Academy of Science, Monograph 3.

———. 2001. *Amphibians and Reptiles of Indiana*. 2nd ed. Indianapolis: Indiana Academy of Science.

Moll, E. O. 1973. "Latitudinal and Intersubspecific Variation in Reproduction of the Painted Turtle, *Chrysemys picta*." *Herpetologica* 29: 307–18.

Monstad, Y. A, and R. Baker. 2004. "Minnesota Frog and Toad Survey 1994–2004." Nongame Wildlife Program. St. Paul: Minnesota Department of Natural Resources.

Moriarty, J. J. 1985a. "Amphibians and Reptiles of the Missouri Drainage of Southwestern Minnesota." Minnesota Herpetological Society Occasional Paper 1: 2–12.

———. 1985b. "Distribution Maps for Reptiles and Amphibians of Minnesota." Minneapolis: Minnesota Herpetological Society.

———. 1986. "A Survey of the Amphibians and Reptiles in South-Eastern Minnesota." Minnesota Herpetological Society Occasional Paper 1: 66–80.

———. 1988. "Minnesota County Biological Survey: 1988 Herpetological Surveys." Minnesota Department of Natural Resources Biological Report Series No. 9.

———. 1990. "Snapping Turtle Bites Off More Than It Can Chew." *Minnesota Herpetological Society News* 10 (6): 4.

———. 1991. "Reintroduction of Bullsnakes into the Crow-Hassan Prairie Restoration." Report to the Nongame Wildlife Program. St. Paul: Minnesota Department of Natural Resources.

———. 1998a. "Blanding's Turtle Workshop Abstracts." Bell Museum of Natural History, Minneapolis.

———. 1998b. "Minnesota Frog and Toad Survey: Results of a Pilot Program." In *Minnesota's Amphibians and Reptiles: Their Conservation and Status,* ed. J. J. Moriarty and D. Jones, 68–71. Lanesboro, Minn.: Serpent's Tales Books.

———. 1999. "Snakes and People: A Guide for Dealing with Snakes in Your Home and Yard." Minnesota Nongame Wildlife Program. St. Paul: Minnesota Department of Natural Resources.

———. 2000. "Blanding's Turtle Workshop Summary." *Chelonian Conservation and Biology* 3: 555–56.

———. 2004. "A Bibliography of Minnesota Herpetology through 2003." Minnesota Herpetological Society Occasional Paper Number 6.

———. 2006. "Natural History Note, Reproduction, Twinning, *Glyptemys insculpta*." *Herpetological Review* 37: 456.

———. 2007. "Bibliography of Minnesota Herpetology: Update 2004 through 2006." Supplement of the *Newsletter of the Minnesota Herpetological Society* 27: 1–8.

Moriarty, J. J., A. Forbes, and D. Jones. 1998. "Geographic Distribution: *Acris crepitans blanchardi.*" *Herpetological Review* 29: 172.

Moriarty, J. J., and D. G. Jones. 1988. "An Annotated Bibliography of Minnesota Herpetology, 1900–1985." Minneapolis: Bell Museum of Natural History.

Moriarty, J. J., and D. Jones. 1997. *Minnesota's Amphibians and Reptiles: Their Conservation and Status*. Lanesboro, Minn.: Serpent's Tale Books.

Moriarty, J. J., and M. Linck. 1995. "Natural History Note. *Emydoidea blandingii*. Reproduction." *Herpetological Review* 26: 99.

———. 1997. "Reintroduction of Bullsnakes into a Recreated Prairie." In *Minnesota's Amphibians and Reptiles: Their Conservation and Status,* ed. J. J. Moriarty and D. Jones, 43–52. Lanesboro, Minn.: Serpent's Tale Books.

Moriarty Lemmon, E., A. R. Lemmon, and D. C. Cannatella. 2007. "Geological and

Climatic Forces Driving Speciation in the Continentally Distributed Trilling Chorus Frogs *(Pseudacris).*" *Evolution* 61: 2086–103.

Munro, D. F. 1949. "Gain in Size and Weight of *Heterodon* Eggs during Incubation." *Herpetologica* 5: 133–34.

Naber, J. R., M. J. Majeski, and A. R. DeMars. 2004. "Baseline Surveys for the Massasauga Rattlesnake in Minnesota, 2002 and 2003." Report to the Nongame Wildlife Program. St. Paul: Minnesota Department of Natural Resources.

Nagel, J. W. 1977. "Life History of the Red-backed Salamander, *Plethodon cinereus,* in Northeastern Tennessee." *Herpetologica* 33: 13–18.

Nelson, W. F. 1963. "Natural History of the Northern Prairie Skink, *Eumeces septentrionalis septentrionalis* (Baird)." Ph.D. diss., University of Minnesota, Minneapolis.

Nyman, S. 1991. "Ecological Aspects of Syntopic Larvae of *Ambystoma maculatum* and the *Ambystoma laterale-jeffersonianum* Complex in New Jersey." *Journal of Herpetology* 25: 505–9.

Oldfield, B. 1988. "Wood Turtle Nesting Study of Lower Cannon River, Goodhue County, Minnesota." Unpublished report to the Nongame Wildlife Program. St. Paul: Minnesota Department of Natural Resources.

Oldfield, B. L., and D. E. Keyler. 1989. "Survey of Timber Rattlesnake *(Crotalus horridus)* Distribution along the Mississippi River in Western Wisconsin." *Wisconsin Academy of Science Transactions* 77: 27–34.

Oldfield, B., and J. J. Moriarty. 1994. *Amphibians and Reptiles Native to Minnesota.* Minneapolis: University of Minnesota Press.

Oliver, J. A. 1955. *The Natural History of North American Amphibians and Reptiles.* Princeton, N.J.: Van Nostrand.

Pace, A. E. 1974. "Systematic and Biological Studies of the Leopard Frogs *(Rana*

pipiens Complex) of the United States." Miscellaneous Publications of the Museum of Zoology, University of Michigan 148: 1–140.

Pallas, D. C. 1960. "Observations on a Nesting of the Wood Turtle, *Clemmys insculpta.*" *Copeia* 1960: 155–56.

Pappas, M. J., and J. Congdon. 2002. "Weaver Bottoms 2001–2002 Turtle Survey: Management and Conservation Concerns." Report to the Nongame Wildlife Program. St. Paul: Minnesota Department of Natural Resources.

Pappas, M. J., J. Congdon, and J. Capps. 2003 (2004). "Turtles Nesting Survey and Salvage of Turtle Eggs in 2003. Response to U.S. Corps of Engineers Management of Dredge Spoil Island #42." Report to the Nongame Wildlife Program. St. Paul: Minnesota Department of Natural Resources.

Perry, P. S., and M. H. Dexter. 1989. "Snakes and Lizards of Minnesota." Nongame Wildlife Program. St. Paul: Minnesota Department of Natural Resources.

Peterson, H. W. 1956. "A Record of Viviparity in a Normally Oviparous Snake." *Herpetologica* 12: 152.

Petranka, J. W. 1998. *Salamanders of the United States and Canada.* Washington, D.C.: Smithsonian Institution Press.

Petranka, J. W., and L. Hayes. 1998. "Chemically Mediated Avoidance of a Predatory Odonate *(Anax junius)* by American Toad *(Bufo americanus)* and Wood Frog *(Rana sylvatica)* Tadpoles." *Behavioral Ecology and Sociobiology* 42: 263–71.

Pettus, D., and G. M. Angleton. 1967. "Comparative Reproductive Biology of Montane and Piedmont Chorus Frogs." *Evolution* 21: 500–507.

Pfingsten, R. A., and F. L. Downs. 1989. *Salamanders of Ohio.* Bulletin of the Ohio Biological Survey, New Series, vol. 7.

Phillips, C. A., and J. Mui. 2005. "Unisexual Members of the *Ambystoma jeffersonianum*

Complex." In *Amphibian Declines: The Conservation Status of United States Species,* ed. M. Lannoo, 640–42. Berkeley: University of California Press.

Phillips, C. A., and O. J. Sexton. 1989. "Orientation and Sexual Differences during Breeding Migrations of the Spotted Salamander, *Ambystoma maculatum*." *Copeia* 1989: 17–22.

Piepgras, S. A. 1998. "Summer and Seasonal Movements and Habitats, Home Ranges and Buffer Zones of a Central Minnesota Population of Blanding's Turtles." Master's thesis, University of North Dakota, Grand Forks.

Piepgras, S., T. Sajwaj, M. Hamernick, and J. Lang. 1998. "Blanding's Turtle *(Emydoidea blandingii)* in the Brainerd/Baxter Region: Population Status, Distribution and Management Recommendations." Final Report to the Nongame Wildlife Program. St. Paul: Minnesota Department of Natural Resources.

Platt, D. R. 1969. "Natural History of the Hognose Snakes *Heterodon platyrhinos* and *Heterodon nasicus*." University of Kansas Publications of the Museum of Natural History 18: 253–420.

Plummer, M. V. 1977. "Activity, Habitat, and Population Structure in the Turtle, *Trionyx muticus*." *Copeia* 1977: 431–40.

Plummer, M. V., and J. C. Burnley. 1997. "Behavior, Hibernacula, and Thermal Relations of Softshell Turtles *(Trionyx spiniferus)* Overwintering in a Small Stream." *Chelonian Conservation and Biology* 2: 489–93.

Plummer, M. V., and D. B. Farrar. 1981. "Sexual Dietary Differences in a Population of *Trionyx muticus*." *Journal of Herpetology* 15: 175–79.

Pope, C. H. 1939. *Turtles of the United States and Canada.* New York: Knopf.

Preston, W. B. 1982. *Amphibians and Reptiles of Manitoba.* Winnipeg: Manitoba Museum of Man and Nature.

Ralin, D. B. 1968. "Ecological and Reproductive Differentiation in the Cryptic Species of the *Hyla versicolor* Complex (Hylidae)." *Southwestern Naturalist* 13: 283–300.

Reeder, T., H. Dessauer, and C. Coles. 2002. "Phylogenetic Relationships of Whiptail Lizards of the Genus *Cnemidophorus* (Squamata: Teiidae)." *American Museum Novitates* 3365: 1–61.

Reigle, N. J. 1967. "The Occurrence of *Necturus* in the Deeper Waters of Green Bay." *Herpetologica* 23: 232–33.

Reinert, H. K. 1981. "Reproduction by the Massasauga *(Sistrurus catenatus catenatus).*" *American Midland Naturalist* 105: 393–95.

Reinert, H. K., D. Cundall, and L. M. Bushar. 1984. "Foraging Behavior of the Timber Rattlesnake, *Crotalus horridus*." *Copeia* 1984: 976–81.

Reinert, H. K., and W. R. Kodrich. 1982. "Movements and Habitat Utilization by the Massasauga, *Sistrurus catenatus catenatus*." *Journal of Herpetology* 16: 162–71.

Rhen, T., and J. W. Lang. 1998. "Among-Family Variation for Environmental Sex Determination in Reptiles." *Evolution* 52: 1514–20.

Roberts, W., and V. Lewin. 1979. "Habitat Utilization and Population Densities of the Amphibians of Northeastern Alberta." *Canadian Field-Naturalist* 93: 144–54.

Rodriguez, E. M., T. Gamble, M. V. Hirt, and S. Cotner. 2009. "Presence of *Batrachochytrium dendrobatidis* at the Headwaters of the Mississippi River, Itasca State Park, Minnesota, USA." *Herpetological Review* 40: 48–50.

Rosenberry, D. O. 2001. "Malformed Frogs in Minnesota: An Update." U.S. Geological Survey Fact Sheet 043-01. U.S. Department of the Interior.

Ross, P., Jr., and D. Crews. 1977. "Influence of the Seminal Plug on Mating Behavior in the Garter Snake." *Nature* 267: 344–45.

Rossman, D. A., N. B. Ford, and R. A. Seigel. 1996. *The Garter Snakes: Evolution and Ecology.* Norman: University of Oklahoma Press.

Rossman, D. A., and P. A. Myer. 1990. "Behavioral and Morphological Adaptations for Snail Extraction in the North American Brown Snakes (Genus *Storeria*)." *Journal of Herpetology* 24: 434–38.

Russell, R. W., G. J. Lipps Jr., S. J. Hecnar, and G. D. Haffner. 2002. "Persistent Organic Pollutants in Blanchard's Cricket Frogs (*Acris crepitans blanchardi*) from Ohio." *Ohio Journal of Science* 102 (5): 119–22.

Sadinski, W., and M. Roth. 2009. "Surveys of Amphibians, Abnormalities, Pathogens, Triazines, Breeding-Site Characteristics, and Reptiles in Five Areas Managed by the National Park Service and the U.S. Fish and Wildlife Service in the Upper Midwest, 2002–2007." NPS Natural Resource Technical Report NPS/GLKN/ NRTR—2009/179.

Sajwaj, T. 1998. "Seasonal and Daily Patterns of Body Temperature and Thermal Behavior in a Central Minnesota Population of Blanding's Turtles (*Emydoidea blandingii*)." Master's thesis, University of North Dakota, Grand Forks.

Savage, W. K., and K. R. Zamudio. 2005. "*Ambystoma maculatum* Shaw, 1802 Spotted Salamander." In *Amphibian Declines: The Conservation Status of United States Species,* ed. M. Lannoo, 621–27. Berkeley: University of California Press.

Sayler, A. 1966. "The Reproductive Ecology of the Red-backed Salamander, *Plethodon cinereus*, in Maryland." *Copeia* 1966: 183–93.

Schaaf, R. T., and J. S. Garton. 1970. "Raccoon Predation on the American Toad, *Bufo americanus*." *Herpetologica* 26: 334–35.

Schloegel, L. M., P. Daszak, A. A. Cunningham, R. Speare, and B. Hill. 2010. "Two Amphibian Diseases, Chytridiomycosis and Ranaviral Disease, Are Now Globally Notifiable to the World Organization for Animal Health (OIE): An Assessment." *Diseases of Aquatic Organisms* 92: 101–8.

Schmid, W. D. 1965. "Some Aspects of the Water Economies of Nine Species of Amphibians." *Ecology* 46: 261–69.

———. 1982. "Survival of Frogs in Low Temperature." *Science* 215: 697–98.

Schroder, R. C. 1950. "Hibernation of Blue Racers and Bullsnakes in Western Illinois." *Natural History Miscellania* 75: 1–2.

Schuett, G. W., D. L. Clark, and F. Kraus. 1984. "Feeding Mimicry in the Rattlesnake *Sistrurus catenatus*, with Comments on the Evolution of the Rattle." *Animal Behavior* 32: 625–26.

Seibert, H. C., and C. W. Hagen. 1947. "Studies on a Population of Snakes in Illinois." *Copeia* 1947: 6–22.

Seigel, R. A. 1986. "Ecology and Conservation of an Endangered Rattlesnake, *Sistrurus catenatus*, in Missouri, U.S.A." *Biological Conservation* 35: 333–46.

Semlitsch, R. D. 1998. "Biological Delineation of Terrestrial Buffer Zones for Pond-Breeding Salamanders." *Conservation Biology* 12: 1113–19.

Sessions, S. K. 2003. "What Is Causing Deformed Amphibians?" in *Amphibian Conservation,* ed. R. D. Semlitch, 168–86. Washington, D.C.: Smithsonian Press.

Sexton, O. J., J. Bizer, D. Gayou, P. Freiling, and M. Moutseous. *Field Studies of Breeding Spotted Salamanders,* Ambystoma maculatum, *in Eastern Missouri, USA.* Milwaukee Public Museum, Contributions in Biology and Geology , no. 67.

Sexton, O. J., C. Phillips, and J. E. Bramble. 1990. "The Effects of Temperature and Precipitation on the Breeding Migration of the Spotted

Salamander *(Ambystoma maculatum)*." *Copeia* 1990: 781–87.

Shaffer, H. B., and M. L. McKnight. 1996. "The Polytypic Species Revisited: Genetic Differentiation and Molecular Phylogenetics of the Tiger Salamander *Ambystoma tigrinum* (Amphibia: Caudata) Complex." *Evolution* 50: 417–33.

Shaw, C. E. 1951. "Male Combat in American Colubrid Snakes with Remarks on Combat in Other Colubrid and Elapid Snakes." *Herpetologica* 7: 149–68.

Shoop, C. R. 1965. "Orientation of *Ambystoma maculatum* Movements to and from Breeding Ponds." *Science* 149: 558–59.

———. 1967. "Relation of Migration and Breeding Activities to Time of Ovulation in *Ambystoma maculatum.*" *Herpetologica* 23: 319–21.

———. 1968. "Migratory Orientation of *Ambystoma maculatum:* Movements near Breeding Ponds and Displacements of Migrating Individuals." *Biological Bulletin* 135: 230–38.

Slevin, J. R. 1951. "A High Birthrate for *Natrix sipedon sipedon* (Linne)." *Herpetologica* 7: 132.

Smith, C. E., and K. H. Kozak. 2011. "A Case of Mistaken Identity: A Re-evaluation of *Pantherophis obsoletus* Distribution in Minnesota." *Newsletter of the Minnesota Herpetological Society* 31 (12): 6–7.

Smith, C. E., and M. B. Zimmer. 2013. "Natural History Notes, *Thamnophis radix,* Brood Size." *Herpetological Review* 44: 454.

Smith, C. K. 1983. "Notes on Breeding Period, Incubation Period, and Egg Masses of *Ambystoma jeffersonianum* (Green) (Amphibia: Caudata) from the Southern Limits of Its Range." *Brimleyana* 9: 135–40.

Smith, G. R., A. Todd, J. E. Rettig, and F. Nelson. 2003. "Microhabitat Selection by Northern Cricket Frogs *(Acris crepitans)* along a West-Central Missouri Creek: Field and Experimental Observations." *Journal of Herpetology* 37: 383–85.

Smith, H. M. 1934. "The Amphibians of Kansas." *American Midland Naturalist* 15: 377–528.

Smith, P. W. 1961. "The Amphibians and Reptiles of Illinois." *Illinois Natural History Survey Bulletin* 28: 1–298.

Smith, W. S. 2008. *Trees and Shrubs of Minnesota.* Minneapolis: University of Minnesota Press.

Somma, L. A. 1987. "Maternal Care of Neonates in the Prairie Skink, *Eumeces septentrionalis.*" *Great Basin Naturalist* 47: 536–37.

Souder, W. 2000. *A Plaque of Frogs.* New York: Hyperion.

———. 2002. Epilogue to *A Plague of Frogs,* 291–300. Minneapolis: University of Minnesota Press.

———. 2005. "Of Men and Deformed Frogs: A Journalist's Lament." In *Amphibian Declines: The Conservation Status of United States Species,* ed. M. Lannoo, 344–48. Berkeley: University of California Press.

Steyermark, A. C., M. S. Finkler, and R. J. Brooks. 2008. *Biology of the Snapping Turtle (Chelydra serpentina).* Baltimore: John Hopkins University Press.

Stiles, R. B. 1938. "The Milk Snakes in Minnesota." *Copeia* 1938: 50.

Storey, K. B., and J. M. Storey. 1986. "Freeze Tolerance and Intolerance as Strategies of Winter Survival in Terrestrially-Hibernating Amphibians." *Comparative Biochemistry and Physiology* A 83: 613–17.

———. 1990. "Frozen and Alive." *Scientific American* 263: 92–97.

Sullivan, B. K., and P. J. Fernandez. 1999. "Breeding Activity, Estimated Age-Structure, and Growth in Sonoran Desert Anurans." *Herpetologica* 55: 334–43.

Swanson, D. L., and S. L. Burdick. 2010. "Overwintering Physiology and Hibernacula Microclimates of Blanchard's Cricket Frogs

at Their Northwestern Range Boundary."
Copeia 2010: 248–54.

Swanson, G. 1935. "A Preliminary List of
Minnesota Amphibians." *Copeia* 1935:
152–54.

Talentino, K. A., and E. Landre. 1991.
"Comparative Development of Two Species of
Sympatric *Ambystoma* Salamanders." *Journal
of Freshwater Ecology* 6: 395–401.

Tanner, W. W., D. L. Fisher, and T. J. Willis. 1971.
"Notes on the Life History of *Ambystoma
tigrinum nebulosum* Hallowell in Utah." *Great
Basin Naturalist* 31: 213–22.

Tattersall, G. J., and P. A. Wright. 1996. "The
Effects of Ambient pH on Nitrogen Excretion
in Early Life Stages of the American Toad
(*Bufo americanus*)." *Comparative Biochemistry
and Physiology* 113: 369–74.

Tester, J. R. 1963. "Techniques for Studying
Movements of Vertebrates in the Field."
In *Radioecology,* ed. V. Schultz and A. W.
Klement Jr., 445–50. New York: Reinhold
Pub. Corp.

———. 1964. "Radio Tracking of Ducks, Deer,
and Toads." *Minnesota Journal of Science* 7:
9–15.

———. 1981. "Growth, Local Movements
and Hibernation of the Manitoba Toad,
Bufo hemiophrys, in Northern Minnesota."
In *Ecology of Reptiles and Amphibians in
Minnesota: Proceedings of a Symposium,
Cass Lake, MN,* ed. L. Elwell, K. Cram, and
C. Johnson, 2.

Tester, J. R., and W. J. Breckenridge. 1964.
"Winter Behavior Patterns of the Manitoba
Toad, *Bufo hemiophrys,* in Northwestern
Minnesota." *Annales Academiae Scientiarum
Fennicae Series. A. IV Biology* 71: 423–31.

Tester, J. R., A. Parker, and D. B. Siniff. 1965.
"Experimental Studies on Habitat Preferences
and Thermoregulation of *Bufo americanus,
B. hemiophrys,* and *B. cognatus.*" *Journal of
Minnesota Academy Science* 33: 27–32.

Timber Rattlesnake Recovery Team. 2008.
"Timber Rattlesnake Recovery Plan."
Division of Ecological Resources. St.
Paul: Minnesota Department of Natural
Resources.

Utiger, U., N. Helfenberger, B. Schaetti,
C. Shmidt, and M. Rus. 2002. "Molecular
Systematics and Phylogeny of Old and New
World Ratsnakes, *Elaphe* auct., and Related
Genera (Reptilia, Squamata, Colubridae)."
Russian Journal of Herpetology 9: 105–24.

Uyehara, I. K., T. Gamble, and S. Cotner. 2010.
"The Presence of *Ranavirus* in Anuran
Populations at Itasca State Park, Minnesota,
USA." *Herpetological Review* 41: 177–79.

Van Buskirk, J., P. Anderwald, S. Lüpold,
L. Reinhardt, and H. Schuler. 2003. "The
Lure Effect, Tadpole Tail Shape, and the
Target of Dragonfly Strikes." *Journal of
Herpetology* 37: 420–24.

Van Gorp, C. D. 1996. "Survey for Blanchard's
Cricket Frog (*Acris crepitans blanchardi*) in
Southwestern Minnesota." Final Report to
the Natural Heritage Program and Nongame
Wildlife Program. St. Paul: Minnesota
Department of Natural Resources.

Van Gorp, C. D., and T. J. VanDeWalle. 1995.
"Survey for Blanchard's Cricket Frog
(*Acris crepitans blanchardi*) in Southeastern
Minnesota." Report to the Minnesota County
Biological Survey. St. Paul: Minnesota
Department of Natural Resources.

Verrell, P. A. 1982a. "Male Newts Prefer Large
Females as Mates." *Animal Behavior* 30:
1254–55.

———. 1982b. "The Sexual Behavior of the
Red-spotted Newt *Notophthalmus viridescens*
(Amphibia: Urodela: Salamandridae)."
Animal Behavior 30: 1224–36.

Vogt, R. C. 1980. "Natural History of the Map
Turtles *Graptemys pseudogeographica* and
G. ouachitensis in Wisconsin." *Tulane Studies
in Zoology and Botany* 22: 17–48.

———. 1981. *Natural History of Amphibians and Reptiles of Wisconsin*. Milwaukee Public Museum, Wisconsin.

———. 1983. "Systematics of the False Map Turtles (*Graptemys pseudogeographica* complex: Reptilia, Testudines, Emyidae)." *Annals of Carnegie Museum* 62: 1–46.

Waldman, B. 1982. "Adaptive Significance of Communal Oviposition in Wood Frogs (*Rana sylvatica*)." *Behavioral Ecology and Sociobiology* 10: 169–74.

Walker, C. F. 1946. "Amphibians of Ohio, Part I: The Frogs and Toads." *Ohio State Museum Science Bulletin* 1 (3): 1–109.

Weed, A. C. 1922. "New Frogs from Minnesota." *Proceedings of the Biological Society of Washington* 35: 107–10.

Wells, K. D. 1977. "Territoriality and Male Mating Success in the Green Frog (*Rana clamitans*)." *Ecology* 58: 750–62.

———. 2007. *The Ecology and Behavior of Amphibians*. Chicago: University of Chicago Press.

Wendt, K. M., and B. A. Coffin. 1988. "Natural Vegetation of Minnesota: At the Time of the Public Land Survey 1847–1907." Minnesota Department of Natural Resources Biology Report. No. 1.

Westerveld, Jay. 2012. "Tiny Caller: The Northern Cricket Frog." *New York State Conservationist* 66 (4): 24–27.

Whitaker, J. O. 1971. "A Study of the Western Chorus Frog, *Pseudacris triseriata,* in Vigo County." *Indiana Journal of Herpetology* 5: 127–50.

Whitford, P. C. 1991. "Final Report on Blanchard's Cricket Frog Survey of Southeastern Minnesota: 1990/1991." Report to the Nongame Wildlife Program. St. Paul: Minnesota Department of Natural Resources.

Whitford, W. G., and A. Vinegar. 1966. "Homing, Survivorship, and Overwintering of Larvae in Spotted Salamanders, *Ambystoma maculatum*." *Copeia* 1966: 515–19.

Wilbur, H. M., and J. P. Collins. 1973. "Ecological Aspects of Amphibian Metamorphosis." *Science* 182: 1305–14.

Williams, K. L. 1978. "Systematics and Natural History of the American Milksnake, *Lampropeltis triangulum*." Milwaukee Public Museum *Publications in Biology and Geology* 2: 1–258.

Williams, P. K. 1969. "Ecology of *Bufo hemiophrys* and *B. americanus* Tadpoles in Northwestern Minnesota." Master's thesis, University of Minnesota, Minneapolis.

Winchell, N. H. 1911. *The Aborigines of Minnesota*. St. Paul: Minnesota Historical Society.

Woodward, J. E., and M. W. Meyer. 2003. "Impact of Lakeshore Development on Green Frog Abundance." *Biological Conservation* 110: 277–84.

Woolverton, E. 1961. "Winter Survival of Hatchling Painted Turtles in Northern Minnesota." *Copeia* 1961: 109.

Wright, A. H. 1914. *North American Anura: Life Histories of the Anura of Ithaca, New York*. Carnegie Institution Publication 197.

———. 1920. *Frogs: Their Natural History and Utilization*. U.S. Bureau of Fisheries Doc. 888.

Wright, A. H., and A. A. Wright. 1949. *Handbook of Frogs and Toads*. Ithaca, N.Y.: Comstock.

———. 1957. *Handbook of Snakes*. 2 vols. Ithaca, N.Y.: Comstock.

Yurewicz, K. L., and H. M. Wilbur. 2004. "Resource Availability and Costs of Reproduction in the Salamander *Plethodon cinereus*." *Copeia* 2004: 28–36.

Index

John J. Moriarty is senior wildlife manager for the Three Rivers Park District located west of Minneapolis and St. Paul. He previously worked as natural resources manager for Ramsey County Parks. He is coauthor with Barney Oldfield of *Amphibians and Reptiles Native to Minnesota* (Minnesota, 1994) and author of *Turtles and Turtle Watching in North Central States.*

Carol D. Hall has coordinated amphibian and reptile surveys for the Minnesota Biological Survey with the Minnesota Department of Natural Resources since 1991. She previously worked with the U.S. Fish and Wildlife Service and the Minnesota chapter of the Nature Conservancy.

Carrol L. Henderson is Nongame Wildlife Program supervisor at the Minnesota Department of Natural Resources.